Pectins - Extraction, Purification, Characterization and Applications

Edited by Martin Masuelli

Published in London, United Kingdom

IntechOpen

Supporting open minds since 2005

Pectins - Extraction, Purification, Characterization and Applications
http://dx.doi.org/10.5772/intechopen.78880
Edited by Martin Masuelli

Part of IntechOpen Book Series: Biochemistry, Volume 6
Book Series Editor: Miroslav Blumenberg

Contributors
Rudy Situmeang, Xinyue Zhang, Yanzhi He, Zhiping Zhang, Erumalla Venkatanagaraju, N Bharathi, Rachiraju Hema Sindhuja, Rajshree Roy Chowdhury, Yarram Sreelekha, Mahejibin Khan, Vanitha Thiraviam, Virginia Rodríguez, Lucía Isabel Castro Vazquez, Belén Ayestarán, Leticia Martínez-Lapuente, Zenaida Guadalupe, Martin Alberto Masuelli

First published in London, United Kingdom, 2020 by IntechOpen
IntechOpen is the global imprint of INTECHOPEN LIMITED, registered in England and Wales, registration number: 11086078, 7th floor, 10 Lower Thames Street, London, EC3R 6AF, United Kingdom
Printed in Croatia

British Library Cataloguing-in-Publication Data
A catalogue record for this book is available from the British Library

Additional hard and PDF copies can be obtained from orders@intechopen.com

Pectins - Extraction, Purification, Characterization and Applications
Edited by Martin Masuelli
p. cm.
Print ISBN 978-1-78984-071-1
Online ISBN 978-1-78984-072-8
eBook (PDF) ISBN 978-1-83968-550-7
ISSN 2632-0983

For EU product safety concerns:
IN TECH d.o.o., Prolaz Marije Krucifikse Kozulić 3, 51000 Rijeka, Croatia, info@intechopen.com or visit our website at intechopen.com.

IntechOpen Book Series
Biochemistry
Volume 6

Martin Masuelli is a Inv. Adj. of the Instituto de Física Aplicada – CONICET – National University of San Luis (UNSL) and Associate Professor, UNSL, San Luis, Argentina. He holds a Master's degree and a PhD and Master thesis in Membrane Technology from UNSL. He is the Director of Physics Chemistry Service Laboratory, UNSL. He is an expert in polysaccharides and physics chemistry of macromolecules. He is the author or co-author of more than 30 peer-reviewed international publications, eight book chapters, 70 communications in international congresses, and editor of seven books. He is a member of the Sociedad Argentina de Ciencia y Tecnología Ambiental, Asociación Argentina de Fisicoquímica y Química Inorgánica and Asociación Argentina de Tecnólogos de Alimentos. From July 2013, he has been the Editor-in-chief and Founder of the Journal of Polymer and Biopolymers Physics Chemistry, Science and Education Publishing. He is on the editorial boards of numerous journals.

Editor of Volume 6: Martin Masuelli
Department of Chemistry; Faculty of Chemistry, Biochemistry and Pharmacy; National University of San Luis, Argentina

Book Series Editor: Miroslav Blumenberg
NYU Langone Medical Center, New York, USA

Scope of the Series

Biochemistry, the study of chemical transformations occurring within living organisms, impacts all of life sciences, from molecular crystallography and genetics, to ecology, medicine and population biology. Biochemistry studies macromolecules - proteins, nucleic acids, carbohydrates and lipids –their building blocks, structures, functions and interactions. Much of biochemistry is devoted to enzymes, proteins that catalyze chemical reactions, enzyme structures, mechanisms of action and their roles within cells. Biochemistry also studies small signaling molecules, coenzymes, inhibitors, vitamins and hormones, which play roles in the life process. Biochemical experimentation, besides coopting the methods of classical chemistry, e.g., chromatography, adopted new techniques, e.g., X-ray diffraction, electron microscopy, NMR, radioisotopes, and developed sophisticated microbial genetic tools, e.g., auxotroph mutants and their revertants, fermentation etc. More recently, biochemistry embraced the 'big data' omics systems.

Initial biochemical studies have been exclusively analytic: dissecting, purifying and examining individual components of a biological system; in exemplary words of Efraim Racker, (1913 –1991) “Don’t waste clean thinking on dirty enzymes.” Today however, biochemistry is becoming more agglomerative and comprehensive, setting out to integrate and describe fully a particular biological system. The ‘big data’ metabolomics can define the complement of small molecules, e.g., in a soil or biofilm sample; proteomics can distinguish all the proteins comprising e.g., serum; metagenomics can identify all the genes in a complex environment e.g., bovine rumen. This Biochemistry Series will address both the current research on biomolecules, and the emerging trends with great promise.

Contents

Preface

Pectin is one of the main constituents of the cell wall of vegetables and it is an important part of the characteristic components of citrus fruits. These macromolecules are highly hydrophilic polysaccharides that can absorb several times their weight in water. Pectins are obtained mainly from citrus fruits. The pectin industry is one of the most important worldwide, where the products generated are for human consumption. Pectin is best known as a gelling agent, as a texturizer, emulsifier, thickener, and stabilizer. This biopolymeric extract, unlike others, develops a gel in the presence of water, sugar, and acid. The main application of pectins in the food industry is in the manufacture of sweets, jellies, and jams. It is also used as a gelling agent in puddings, as a stabilizer of emulsions and suspensions, a thickening agent in drinks, a stabilizing agent in ice cream and cold desserts, and in solutions for coating sausages and canned meats. In the pharmaceutical field, pectins are used for their protective and regulatory action in the gastrointestinal system. Their detoxifying, anti-cholesterol, immunological, anti-hemorrhagic, anti-cancer and healing effects prolong the therapeutic action by increasing the release times of the active ingredients. It is also used in the formation of films for coating paper and giving softness characteristics in wrapping paper, and as a vehicle in the preparation of barium sulfate suspensions for application in X-ray radiographs. As substitutes for petroleum derivatives, mixtures of pectin and polyvinyl alcohol are used in the manufacture of biodegradable films. These films are biodegradable, recyclable, and allowed for prolonged release of pharmaceutical forms and as protectors or adhesives in pharmaceutical preparations. The pectin industry is a very important engine in the global economy of biopolymers, mainly because of its availability and abundance given in citrus farming.

This book aims to study the use of pectins obtained from citrus and other sources, to obtain food and pharmaceutical additives.

This book consists of the following sections and chapters:

Section 1: Pectin Extractions

Chapter 1: **Extraction and Characterization of Pectins from Peels of Criolla Oranges (*Citrus sinensis*). Experimental Reviews.** Authored by P. Ruano, L. Lazo Delgado, S. Picco, L. Villegas, F. Tonelli, M. E. Aguilera Merlo, J. Rigau, D. Diaz, M. Masuelli.

In this chapter the authors study pectin extraction of criolla orange peel (*Citrus sinensis*) by basic or acid hydrolysis and purification and clarification. Once it is purified, it is characterized in aqueous solution to determine its physicochemical properties such as molecular weight, hydrodynamic radius, hydration value, shape factor, etc. Finally, the product is characterized as a film forming with mechanical and thermal tests.

Chapter 2: **Extraction and Purification of Pectin from Agro-Industrial Wastes.** Authored by E. Venkatanagaraju, N. Bharathi, R. H. Sindhuja, R. R. Chowdhury, Y. Sreelekha.

This chapter focuses on extraction and purification of pectin from various agro-industrial wastes, considered to be main environmental pollutants. The authors explore other sources of pectin or modify the existing sources to obtain pectin with the desired quality attributes. The large variety of applications as well as the increasing number of studies on pectin suggests that the potential of pectin as a novel and versatile biomaterial will be even more significant in the future.

Section 2: Pectin Applications

Chapter 3: **Pectin - Extraction, Purification, Characterization and Applications.** Authored by V. Rodríguez Robledo & L. I. Castro Vázquez.

The authors evaluated innovative applications of pectins in the food industry (as a gelling, thickening, and stabilizer agent) and the pharmaceutical industry (bioactive components) including biomedical application (biomaterials, drug delivery, tissue engineering, and wound healing).

Chapter 4: **Role of Pectin in Food Processing and Food Packaging.** Authored by T. Vanitha & M. Khan.

The authors researched a new application for pectin as edible films or coating. These films act as a natural barrier for the exchange of moisture, gases, lipids, and volatiles between food and the environment, and protect fruits and vegetable from microbial contamination.

Chapter 5: **Pectins as Emulsifying Agent on the Preparation, Characterization and Photocatalysis of Nano $LaCrO_3$.** Authored by R. Tahan & M. Situmeang.

The authors conducted studies of environmentally friendly materials such as $LaCrO_3$ perovskite with pectin as the emulsifying agent. This material is characterized by the photocatalytic reaction with dyes and cellulose conversion.

Section 3: Pectin Interactions

Chapter 6: **Properties of Wine Polysaccharides.** Authored by L. Martínez-Lapuente, Z. Guadalupe, B. Ayestarán.

This chapter describes the origin, structure, and role of the different wine polysaccharide families through a bibliographic revision of their origin and extraction into the wines, as well as their technological and sensory properties. There are only a few studies regarding their effects and mechanisms of action, and more research needs to be done to better learn about their role and applicability in the wine industry.

Chapter 7: **Flavonoids and Pectins.** Authored by Z. Zhang, Y. He, and X. Zhang.

In this chapter, we present current opinions on flavonoids to understand how to classify this group of secondary metabolites, what biological and pharmaceutical activities they possess, and how to biosynthesize them in plants.

This book is aimed at professionals and researchers of pectins and biopolymers, teachers and students of universities, and industrial and pharmaceutical academics.

Martin Alberto Masuelli
Associate Professor Sup.,
Area of Physical Chemistry,
Department of Chemistry,
Faculty of Chemistry, Biochemistry and Pharmacy,
National University of San Luis,
San Luis, Argentina

Adj. Researcher,
Institute of Applied Physics-CONICET,
San Luis, Argentina

Director of the Physical Chemistry Research and Services Laboratory
(LISEQF-UNSL),
San Luis, Argentina

Section 1

Pectin Extractions

Chapter 1

Extraction and Characterization of Pectins From Peels of Criolla Oranges (*Citrus sinensis*): Experimental Reviews

Paula Ruano, Lismet Lazo Delgado, Sergio Picco, Liliana Villegas, Franco Tonelli, Mario Eduardo Aguilera Merlo, Javier Rigau, Darío Diaz and Martin Masuelli

Abstract

The citriculture in the international field is important not only for its nutrient and vitamin characteristics but also for its valuable source of raw materials to obtain pectin, since it was found in the internal and external parts of the citrus peel. In our country, there are several varieties of orange, from which you can obtain by-products other than juice, such as essential oils, fertilizers, concentrates, and pectin. Pectin has different uses in the pharmaceutical industry for the preparation of suspensions, emulsions, cosmetics, capsules, etc. In the food industry, it is used as a film in packaging, thickener, and gelling agent in the manufacture of jellies and preserves, in wines as dehydrating plant tissues, in milk to precipitate casein, etc. The pectin extraction was carried out by basic or acid hydrolysis and then proceeds to purify and clarify. Once purified, the pectin was characterized in aqueous solution to determine its physicochemical properties such as molecular weight, hydrodynamic radius, hydration value, shape factor, etc. Thermal and mechanical characterizations were also performed, to assess its ability to form films.

Keywords: pectin, *Citrus sinensis*, hydrolysis, films

1. Introduction

Pectin is a structural polysaccharide found in all higher plant fruits as citrus, apples, grapes, plums, etc. and is, therefore, part of the natural man diet. Many aspects of plant physiology and pathology, food texture, and even wine production involve pectin and its fate in materials and organisms. Commercial preparations of pectin are usually derived from citrus or apple peels, by-products of juice manufacture. The production involves aqueous extraction under mild acidic conditions, followed by precipitation by the addition of a di- or trivalent metal alcohol or ions.

Most of the world production of pectin is used for the preparation of jams and jellies, but a growing part is used in confectionery products, beverages, and acidified milk drinks. Pectin is suitable for applications in acidic food products due to its good stability at low pH values [1]. For this same reason, it is currently also used in the pharmaceutical industry as in the manufacture of capsules, films, and biodegradable patches.

Pectins are complex heterogeneous polysaccharides; it is used plurally because it differs from its composition from one species to another (the pectin obtained from a citrus differs, e.g., from the apple; even within citrus, there are small differences). Like most other vegetable polysaccharides, it is both polydispersed and polymolecular, and its composition varies with the source and conditions applied during storage. In any pectin sample, parameters such as weight or content of particular subunits will differ from molecule to final molecule.

All pectin molecules contain linear segments of α (1 → 4)-D-galactopyranosyl uronic acid units linked with some of the carboxyl groups esterified with methanol (**Figure 1**). In the pectin of some sources, some of the hydroxyl groups of the galacturonosyl units (0–2 and/or 0–3, oxygen bonded to carbon 3) are esterified with acetic acid [2].

Pectins are a mixture of acidic and neutral branched polymers. They constitute 15–30% of the dry weight of the primary wall of plant cells. They determine the porosity of the wall and therefore the degree of availability of the substrates of the enzymes involved in the modifications of the same. Pectins also provide charged surfaces that regulate pH and ion balance. In the presence of water, they form gels. Pectins have three main domains:

1.1 Homogalacturonans (HG)

Compounds by D-galacturonic acid (GalU) residues are bound by an α (1 → 4) bond. The carboxyl groups of C6 (carbon number 6 of the GalU) can be methyl esterified or remain free. The free carboxyl groups, if dissociated, give rise to calcium bonds between the neighboring HG chains, forming the so-called egg box structure. For a region of HG to be sensitive to calcium binding, 10 molecules of unesterified GalU are required, the formation of bonds of this type is related to the arrest of the cell wall and, therefore, with the cessation of growth and increase of stiffness of the wall. The GalU can be found acetylated in O2 (oxygen number 2 of the GalU) or in O3 [2].

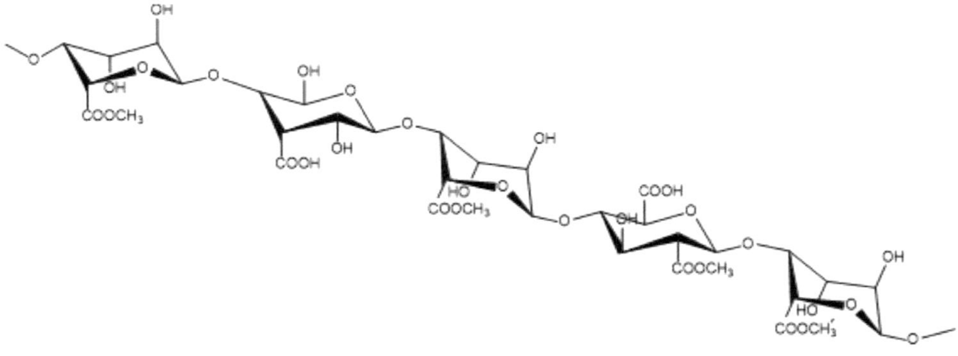

Figure 1.
Basic chemical structure of α (1 → 4)-D-polygalactopyranosyl uronic acid.

1.2 Rhamnogalacturonan-I (RGI)

GalU linked in α (1 → 4) with L-rhamnose (Rha) residues intercalated with an α (1-2) bond, that is, [(1-2)-α-L-Rha-(1-4)-α-D-GalU] n, where n can be greater than 100. These Rha residues are the anchoring of side chains, and approximately half are bound by C4 to chains of arabinans, formed by α-L-arabinose (Ara) linked in α (1 → 5) as the main axis that can be substituted with the chains Ara (1-2)-α-Ara (1-3) and/or Ara (1-3)-α-Ara (1-3) or arabinogalactan I (AGI), chains of β-(1-4)-D-galactose (Gal), with C6-Gal branches. They can also be substituted for α (1 → 5) Ara in Gal C3 [2].

1.3 Rhamnogalacturonan-II (RGII)

Small polysaccharide of very complex structure, formed by GalU, Rha, Ara, Gal, and small amounts of infrequent sugars such as apiose or acetic acid. The Rha moieties may be substituted at C3; in C3 and C4, in C2, C3, and C4 or be terminal. The arabinogalactan of the RGII presents ramifications in C3 and C6 of Gal and in C3 and C5 of Ara. The side chains contain a high number of different residues bound with various bonds. However, the RGII has a highly conserved structure and can form dimers via a borate bridge, with two ester bonds [2].

Arabinans and galactans of the RGII of the *Amaranthaceae* family can be associated to ferulic acid through an ester bond, which makes it possible to link several chains by diferulic bridges, through the action of the peroxidases. Links are also caused by the dimerization of hydroxycinnamic acids linked to arabinans and galactans of the RGI due to the action of peroxidases [2, 3].

Commercial pectins are derived almost exclusively from citrus or apple, both by-products of the manufacture of juice or cider. While the apple pomace contains 10–15% pectin on a dry matter basis, the citrus peel contains 20–30%. The pectin of citrus fruits and apples is equivalent from the point of view of the application. However, citrus pectins are light cream or light tan; apple pectins are often darker.

Other suggested alternative sources include sugar beet residues, sunflower heads (seeds used for edible oil), and mango residues. The pectin from sugar beet was produced in Germany during World War II and in Sweden and Russia in the following years. The beet pectin is inferior to the citric pectin in molecular weight.

All currently significant applications are related to (1) the degree of acetate esterification, (2) relatively low molecular mass, and (3) the presence of large amounts of neutral sugar side chains [4].

The use of pectin in traditional sugar jams is one of the best known applications, being one of the largest pectin markets. Very often the pectin is the only gelling agent allowed, with 0.2–0.4% of pectin, depending on the type and origin of the fruit. Within the European Economic Community, there are two standards in jam and extra marmalade, which contain a minimum of 30 or 45% fruit pulp, respectively. The higher quality of jam also tends to be made with better quality fruit, so it requires significantly less pectin.

The jam of citrus fruit, especially lemon or grapefruit, in which pectin has been produced in a higher than normal content, generates disadvantages. In this case, too much pectin in the fruit leads to too strong gel formation or even to pregelation and syneresis; the attempt to regulate the texture with the pH leads to a situation in which the control is extremely critical and the rejection rate to this pectin is inevitable. One solution has been to work at a pH where the pectin in the fruit no longer forms a gel. The gelation can then be achieved by adding a pectin-amide of low methoxyl content which is capable of gelling at high pH [4].

Both the selection of the correct pectin (the lower the solid solubility, the more sensitive to calcium) and the pectin content in fruit are important. Sometimes, especially at very low amounts of pectin, it may be necessary to add a calcium salt to obtain a better result. From time to time, neutral gums are added to reduce syneresis, but each attempt must be made to optimize the type and level of pectin. The exact pH before considering such addition should be considered, since the gums can mask the delicate taste of these products [4].

Jam makers also manufacture fillings and toppings for the bakery and related industries. Many of these use pectin as a gelling agent or thickener, but, because it depends so much on their processing conditions, it is very difficult to generalize a process on an industrial scale.

In recent years, a growing area of the fruit product industry is the production of fruit for addition to yogurts and similar products. Many of these have been made with modified starch as a thickener to ensure a homogeneous distribution of the fruits together with the texture that must be extruded and the difficulties that this entails. Unfortunately, although starches are relatively inexpensive, they can mask delicate fruit flavors and lead to a mealy texture. These foods have sugar contents between 30 and 60%, so high methoxyl pectin can be used. Pectins have other uses in the dairy industry, for example, the high methoxyl pectin prevents the aggregation of casein in the heating to pH values lower than 4 or 3. It can therefore be used as a stabilizer for the drinkable yogurts treated with UHT and for mixtures of milk and fruit juices. It will also stabilize acidified soy milk drinks and whey products. The yogurt can be thickened by adding very low levels of low pectin in amidated methoxyl before cultivation. Although this is not allowed in many countries, a suitable pectin incorporated in a fruit base can, with careful formulation, have an effect comparable to a fruit yogurt. On the other hand, low-calorie soft drinks are often thin and lack the characteristic mouthfeel provided by sugar in conventional soft drinks. A low level of pectin (usually of controlled viscosity) can be used to improve the texture of these products and also to replace part of the texture due to fruit pulp in juice formulations [5].

1.4 Pectin extraction and purification

For pectin extraction two general processes are used: (1) those that separate pectins from most of the other materials by precipitation with an alcohol and (2) those that precipitate pectins as an insoluble salt with suitable multivalent metal ions (**Figure 2**). Both can be used to obtain any pectin within the two main groups, namely, high methoxyl pectin (HM-pectin) and low methoxyl pectin (LM-pectin).

The conditions chosen for the extraction depend on the raw material and the desired product. Temperatures between 50 and 90 °C, pH 1-3, and with extraction times from 30 minutes to 24 hours. Acidification can be done with sulfuric acid, sulfurous acid, hydrochloric acid or nitric acid, in accord with reference [4, 5].

Processes with a long extraction time, where the pH is low and the temperature is high, are conducive to high product yield, but the quality can be adversely affected. The combination of low temperature with long duration and low pH to obtain some degree of de-esterification (DE) in the extraction in order to produce LM-pectin or on the contrary a slow hydrolysis in time generates HM-pectin [6].

The extraction is followed by filtrations. The raw materials used, which are very soft and swellable, are separated in an initial coarse mesh filtration and sold as livestock feed. The fine suspended solids are subsequently removed from the extract by filtration through diatomaceous earth [7].

If precipitation with alcohol is used to separate the pectin from the extract, the concentration of the extract usually precedes the precipitation, the reason for which

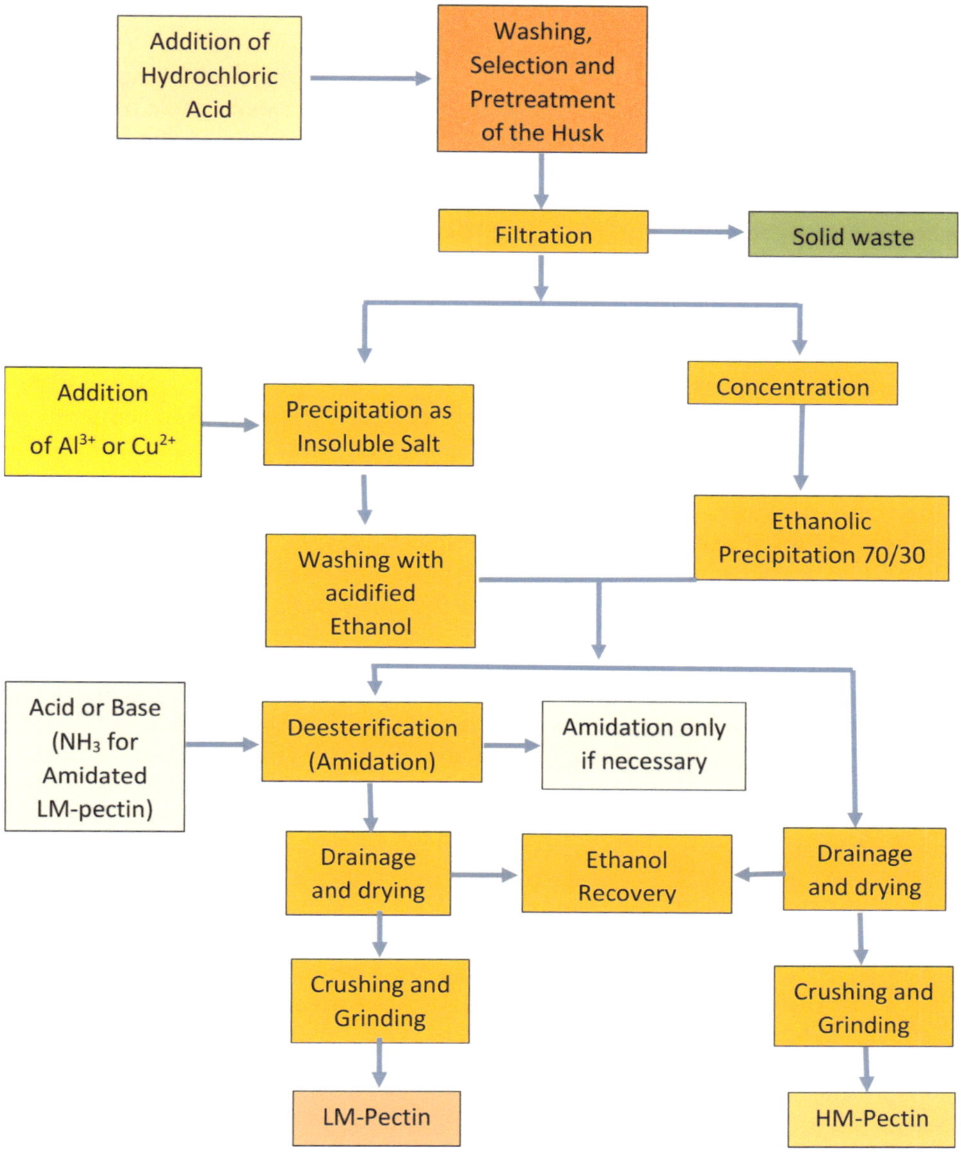

Figure 2.
Manufacturing process of pectins [4].

is saving alcohol. The concentration is usually done by evaporation, but ultrafiltration has been attempted. Al^{3+} is usually chosen if the pectin is separated from the extract as an insoluble one. Precipitation by Cu^{2+} can be an alternative. The metal ions are subsequently removed by several washes with HCl diluted in alcohol [8].

De-esterification to achieve the end of LM-pectin is usually done in an alcohol to which an acid or a base has been added. However, most of all LM-pectin are de-esterified with ammonia, thus producing amidated pectin. The traditional alcohol for the manufacture of pectin is 2-propanol (isopropanol), but also methanol or ethanol can be used. The recovery of alcohol by distillation adds considerable energy costs to production [4, 9].

A simple process of obtaining pectins is the one observed in **Figure 2**, in accordance to Ref. [4].

1.5 Films

A film is a uniform layer that can be formed either from a single component or from a mixture of polymers. The film development with biodegradable polymers is a promising technology in the food packaging industry. Food packaging is used for the preservation and protection of all types of food and its raw materials, particularly oxidative and microbial deterioration, as well as to extend the shelf life. The increased use of synthetic packaging films has led to serious ecological problems because they are not completely biodegradable and recyclable. The reduction of environmental pollution has led to a paradigm shift in the use of biodegradable materials, especially renewable raw materials in agriculture and waste processing industry food. This approach is equivalent to the conservation and recycling of natural resources, as well as the generation of new packaging of innovative design. The total packaging biodegradation generates benign products with the environment, and the permeability of CO_2 and water is a parameter to be considered at the time of the development of the films. Natural polymers cross-linked and copolymerized with synthetic monomers are other alternatives to biodegradable packaging films. For the time being, the complete replacement of synthetic plastics is simply impossible to achieve and may even be unnecessary, at least for some specific applications that require our attention in the future. Definitely, biodegradable bio-packaging will be very promising in the near future [10–14]. For this reason, the development of edible films and coatings has been extensively studied in recent decades. These structures act as a barrier between food and the environment, helping the outer packaging in its protective function. In addition, some of them are supposed to have their own sensory or nutritional characteristics. The use of fruit purees in edible films has been previously explored by conferring distinctive flavors and colors that could be exploited for applications such as sushi wraps, fruit strips, or colorful coatings for specific foods [15].

The applications of biopolymers are from foods (nutritive and dietary fibers), packaging (biodegradable bags and/or protective films), thickening agents, gelling agents, foaming agents, and emulsifiers [16]. As films or membranes, are used in separative processes, for gas separation, diffusion, filtration and reverse osmosis [16]. In the pharmaceutical industry, the biofilms are used as drug encapsulants, films, and patches [17]. On the other hand, in mining industry, they are utilized as flocculating agents and precipitants of heavy metals [18].

The publications of biofilms are framed in the first place in the development of new materials from polysaccharides of great abundance and wood and agroindustrial production such as cellulose acetate [19] and starch [20]. Other polysaccharides available, but at high cost, are alginates [21], chitosan [22], and guar gum [23]. From an original point of view, the development of new films with biopolymers of various types such as arabinoxylans [24], soybean polysaccharides [25], and watercress polysaccharides [26] is worth highlighting. To clarify the latter, the latest advances in these films are described below.

1.5.1 Regarding pectin films

Pectin due to its biodegradability, biocompatibility, edibility, and versatile chemical and physical properties (such as gelling, selective gas permeability, etc.) is a polymer matrix suitable for the production of edible films intended for the packaging of active foods. It is understood by an edible film as a packaging material, which is a thin layer of edible material placed on or between food components. Active packaging is a system that has basic barrier functions and others that are achieved by incorporating active ingredients in the packaging material and/or by using functionally active polymers. For example, if the packaging system has antimicrobial activity, packaging limits microbial growth by extending the latency period. Espitia et al. [27] performed an interesting review that describes the main methods for manufacturing edible pectin films, the main characterization techniques to determine their physico-mechanical properties, and the applications of edible pectin films as antimicrobial food packaging.

Parris et al. [28] evaluated the water vapor permeability of hydrophilic films of alginate and pectin. Their results suggested that these diminished the mechanical properties by incorporating whole milk, sodium caseinate, skimmed milk powder, or whey in the film. In addition, this work evaluated the appropriate choice of plasticizer. Sodium alginate films exhibited lower water vapor permeability values than films prepared using low or high methoxylation pectin. The author found that sodium lactate was an effective plasticizer and alginate films containing 50% by weight or more of sodium lactate had an elongation greater than 13%. Films prepared with sorbitol as a plasticizer had the best water vapor permeability values but tended to be stiff and in some cases too fragile for tensile measurements. The addition of mixtures of whole milk to film effectively reduced water vapor permeability values by up to 35%.

On the other hand, Pasini Cabello et al. [29] studied the effect of two plasticizers, glycerol (GLY) and polyethylene glycol (PEG), on the structure of the pectin films. The results revealed that glycerol acted as an internal plasticizer. Meanwhile, glycerin increased the predominant amorphous character of the plasticized films due to the decrease in intermolecular attraction, which resulted in degradation at low temperature and allowed the conformational transformation of the film to galacturonan ring through a can conformation. Glycerol produced more deformable and weaker films. In addition, glycerol produced films with a higher swelling index (SI) and a water vapor permeability value (WVP). When PEG was used as a plasticizer, a lower Young's modulus was obtained than the pure pectin film. However, by increasing the molecular weight of polyethylene glycol, more compact and less deformable films were obtained. WAXD spectra and DSC thermograms indicate that PEG works as a separate phase in the pectin matrix, more compact and less permeable to water vapor as the molecular weight of PEG increases. These results show that PEG acts as an external plasticizer.

Azeredo et al. [30] conducted the investigation of pomegranate juice into pectin films, giving it a bright red color, and also acted as a plasticizer. The increase in the pomegranate juice/water ratio from 0/100 to 100/0 resulted in an improved elongation (from 2 to 20%), a decrease in strength (from 10 to 2 MPa) and a Young's modulus (from 93 to <10 MPa), an increase in water vapor permeability (WVP, from 3 to 9 g mm $kPa^{-1} h^{-1} m^{-2}$), and a decrease in insoluble matter (IM, of 35–24%). Although an effect of cross-linking (cross-linking) by citric acid was not confirmed, it is demonstrated by its effects on the films. Citric acid markedly increased MI (from <10% to almost 40%); in addition, when measured on a dry film basis, the effects of citric acid showed a notable tendency to increase resistance and modulus and to decrease WVP. The citric acid decreased the density of the red color, which suggests a destabilization of the anthocyanins.

Different characterization techniques allow to determine the properties of biopolymers. Some techniques are used in solution and others in film. The data in solution demonstrate the tendency of the biopolymer to interact with the solvent and the conformation it acquires, besides the possible implications of its molecular weight with the ability to form gels. The techniques that provide this important information are gel permeation chromatography [31–34], ultracentrifugation-sedimentation, polarization to light, refractive index, light scattering, osmometry, diffusometry [24], viscosimetry, densimetry [26], and rheology [25, 35]. When the film is already formed, the most common analyses are water adsorption, either as steam or pure water, water vapor permeation, and swelling index, data that provide information about the affinity of water with the biopolymer. The mechanical tests (Young's modulus) help to elucidate the ductility of the material under study. The structural characteristics are determined by FTIR-ATR, diffraction of RX, SEM, and AFM [36], techniques that help to interpret parameters of the intimate nature of the film and the displacements in its signals caused by its chemical structure. The TGA-DTG and DSC data show the stability of the material against thermal changes during synthesis. In short, all these techniques give us information on the possible applications of each biopolymer and select what type of hydrolysis should be performed. Once the biopolymer has been obtained and characterized, permeoselectivity tests are carried out, such as the controlled release of drugs [37], coagulation or heavy metal precipitation [38], adsorption studies [39], gas permeation [40], and food packaging [41, 42].

In present work, pectin extraction from orange peel was carried out through the hydrolysis process; several extraction conditions were evaluated, specifically the pH in aqueous solution performing acid and basic hydrolyses at different concentrations and times but at constant temperature. Then the pectin obtained was characterized by a comparative analysis with commercial pectin with different techniques described below.

2. Pectin extraction fundament

The citriculture in the international field is important not only for its nutrient and vitamin characteristics; it has also become a valuable source of raw materials to obtain the pectin, since it is found in the internal and external parts of the citrus peel. In our country there are several varieties of orange, from which you can obtain other by-products besides juice, such as essential oils, fertilizers, concentrates, and pectin.

The purpose of this work is to exploit the waste of criolla orange peels, which in some cases causes pollution problems in the ecosystem as it is an adequate source for the proliferation of insects and microorganisms that are harmful to human health, in particular the by-products of citrus fruits used in the preparation of juices and marmalade. On the other hand, the industry generates an expense for the elimination of citrus waste if it is not used. Also, the lack of national raw materials for the pharmaceutical and food industry allows us to search for the natural resources that could be exploited. All these cases motivate to investigate the benefits obtained from citrus waste, such as oranges, which is used as an input for the agro-industry in the production of juices mainly, whose process involves a considerable generation of waste such as husks, pulp, and seeds. This has two benefits; on the one hand, it seeks to increase its "added value" with the process of agro industrialization and, on the other, reduce the environmental impact they produce.

In present work, the pectin extraction was carried out from the orange peel of the variety *Citrus sinensis* and through acidic or basic hydrolysis, by means of pH changes and extraction times at a constant temperature, with which it is expected to obtain a higher yield (**Figure 3**).

There are many processes for the extraction of pectin. In general, the raw material undergoes pretreatment that involves cleaning to remove foreign particles, washing to remove sugar and acid, drying, crushing, and storage. The substrate material is heated at reflux for several hours, with stirring, with an acid of known concentration (sulfur, sulfuric, nitric or hydrochloric acid) [43].

The objective of this work was to compare the different reagents used during hydrolysis and compare performance as well as mechanical, physical, and chemical behaviors. Also, the characterization techniques of pectin were in order to determine its quality.

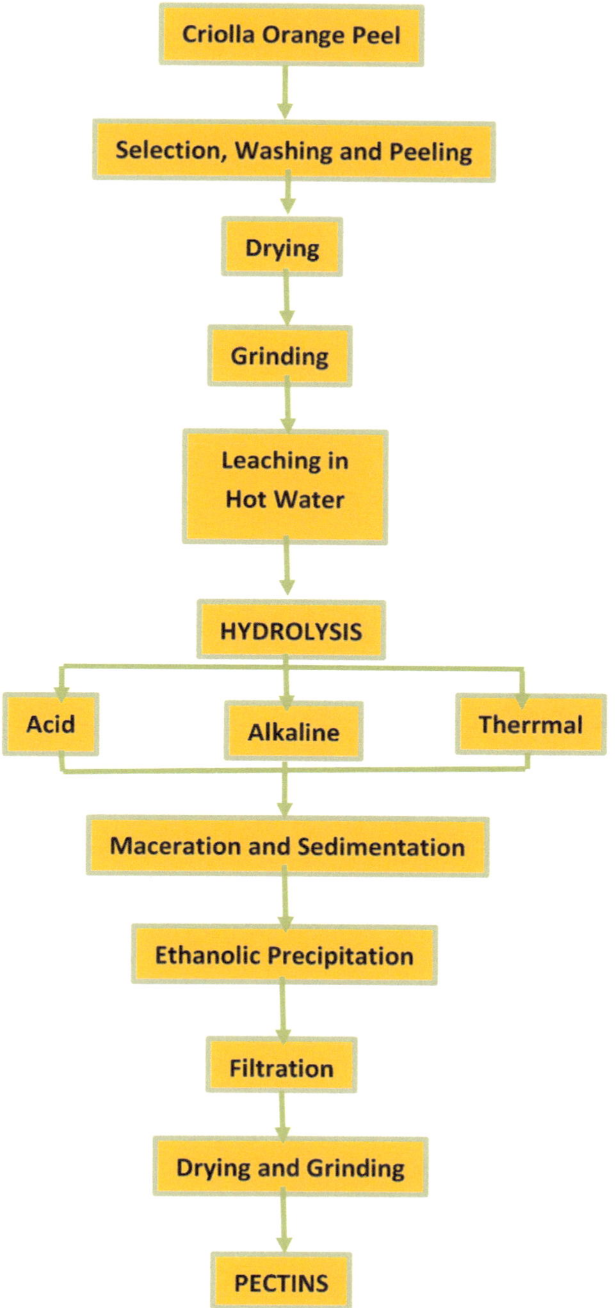

Figure 3.
Scheme of pectin production in this work.

2.1 Pectin extraction from *Citrus sinensis*

Many raw materials and products of the chemical and food Industry require preparation and conditioning. The process begins with the formation of flour from the raw material used, in this case the orange peel. Later, it is treated to achieve the necessary conditions that are required in the subsequent steps to obtaining the final product, the unit operations necessary to carry out the pectin formation process (see **Figures 3–5**).

Selection, washing, and peeling of *Citrus sinensis*: In a first step, the oranges were manually selected. The quantity and quality of useful pectin obtained depend on the species of the fruit, the quantity that the fruit contains naturally, the state of maturation, the management conditions, and the enzymatic activity after harvesting and of the extraction process. They also depend on the part of the fruit that is used. For example, in unripened fruits the greater the amount of pectic material is insoluble in water, the quantity and the solubility increase with maturity [44].

It was decided to investigate as raw material the "criolla," orange which is a very common and easy-to-obtain species, observing that they do not present any type of microorganism or cuts and/or blows that may affect its safety or the development of

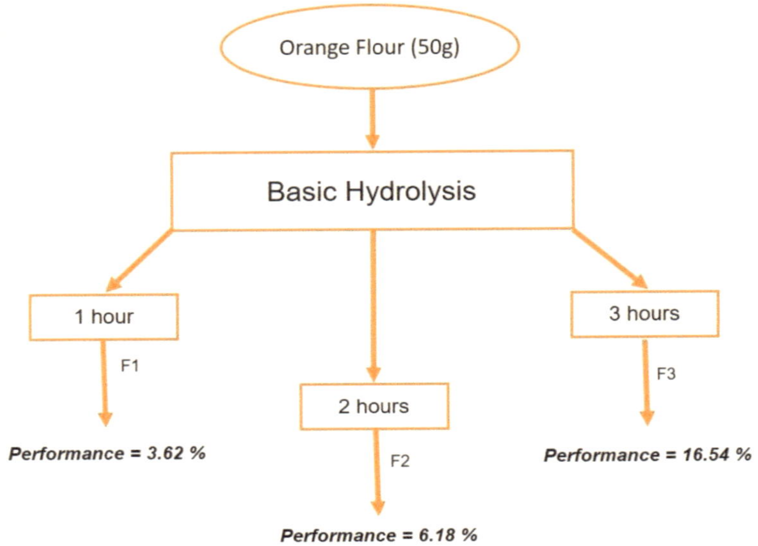

Figure 4.
Representative diagram of F pectins, with their respective performance.

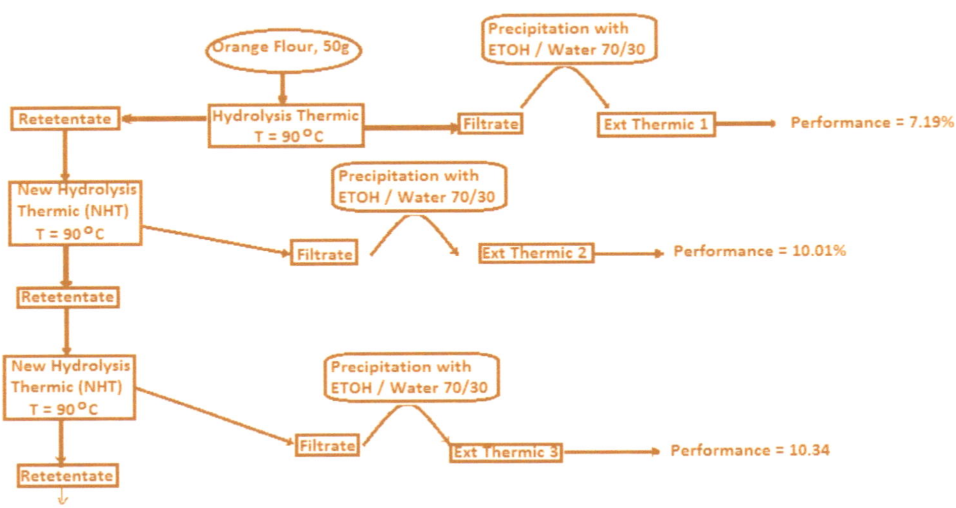

Figure 5.
Representative diagram for thermal hydrolysis with its respective performance.

bacteria or fungi, in addition to being in their commercial maturity. Once selected, the oranges are washed for later use. They are peeled manually without removing the albedo, since it is rich in pectic substances.

2.2 Drying of orange peels

In general, the drying of solids consists in separating small amounts of water or other liquid from a solid material in order to reduce the residual liquid content to an acceptably low value. The liquid to be vaporized can increase on the surface of the solid, as in the drying of saline crystals; inside the solid, as in the case of solvent removal of a sheet of a polymer; or part on the outside and part inside. The feeding of some dryers is a liquid in which the solid is suspended in the form of particles or in solution. The product that dries can withstand high temperatures or requires gentle treatment at low or moderate temperatures. This leads to the existence of a large number of types of commercial dryers in the market [45].

2.3 Pectin obtainment

2.3.1 Hydrolysis

The degradation of the polysaccharides begins, in general, at the reducing end of the molecule and proceeds step by step through the anhydroglucose chain [46].

The degradation of the polysaccharides proceeds by a peeling process in which the reducing end group is released from a chain by removing the remainder of the chain as a glycoxyl anion. The elimination takes place when the chain is in the beta position of a carbonyl group of the final reducing unit [47].

The extraction of pectin can be by acid or aqueous base. The basic extraction process produces a pectin of low degree of esterification (low methoxyl pectin) as a result of the saponification of the ester groups, while the acid extraction process generally produces a pectin with a high degree of esterification (high methoxyl pectin), approximately equal to the naturally occurring degree of esterification (DE). The high pectin has an esterification degree of 50% or more. Low DE and high pectin generally have different uses in food products, since they gel by different mechanisms. Both are sold commercially [48].

In the extraction process with acid and base, the plant material was treated with acid or base at temperatures between 70 and 90°C for a sufficient time to eliminate the desired quantities and the quality of the pectin from the cellulose plant material. The pectin was separated from the reaction mixture by filtration. The pectin is precipitated from the extract juice by specific means, either precipitation with alcohol (ethyl or isopropyl can be used) or salifying with aluminum chloride. The precipitated pectin was separated from the precipitating solution by filtration. The extract obtained consists of those molecules that are soluble under the conditions of pH, time, and temperature used during extraction. The extract was composed of a mixture of pectins of different molecular weights and degrees of esterification. Molecular weights can vary from 100,000 to 200,000, but average molecular weights are more typically 140,000 g/mol [48].

Hydrolyzed pectins were compared against commercial pectin from citrus peel by Sigma (galacturonic acid ≥74.0%, methoxy groups 6.7%).

2.4 Maceration and sedimentation

This mechanism is frequently used in the food industry for the separation of solid particles contained in liquids as well as for the separation of two immiscible liquid phases. The driving force is the difference in density between the two phases [45].

The evolution of the sedimentation of a typical flocculated suspension is as described below. **Figure 6** shows a suspension uniformly distributed in the liquid and ready to sediment. The total depth of the suspension is Z. If there are no sands in the mixture, the first appearance of solids in the bottom of the settler is due to the flocs that originate in the lower part of the mixture. As **Figure 6** shows, these solids, consisting of flocs resting gently on each other, form a layer called zone D. Above area D, another layer is formed, called zone C, which was a transition layer whose content in solids varies from that of the original pulp to that of zone D. Above zone C was zone B, which consisted of a homogeneous suspension of the same concentration as the original pulp. Above zone B is zone A which, if the particles have been totally flocculated, is a clear liquid. In well-flocculated pulps the boundary between zones A and B is clear. If particles remain unglued, zone A is cloudy, and the boundary between zones A and B is unclear. **Figure 6c** shows that as sedimentation progresses, the thicknesses of zones D and A increase, and that of zone C remains constant, while that of zone B decreases.

The whole process represented in **Figure 6** is called sedimentation. Subsequently, as shown in **Figure 6d**, zones B and C disappear, and all the solids are in zone D, beginning then a new effect called compression. The moment in which the compression is evident for the first time is called a critical point. In the compression part of the liquid that accompanied the flocs in the compression, zone D is expelled when the weight of the deposited solids breaks the structure of the flocs. During compression, a part of the liquid contained in the flocs is projected out of zone D forming small jets, and the thickness of this area decreases. Finally, as shown in **Figure 6e**, when the weight of solids reaches the mechanical equilibrium with the resistance to compression of the flocs, the sedimentation process is stopped. At this time the silt reaches its final height [45].

In this stage, the mixture was left to marinate for 24 h so that the unusable solids would settle. The separation of the phases was observed until obtaining the solid at the bottom of the crystallizer and in this way to be able to more efficiently separate the soluble pectin found in the supernatant, illustratively in the area A of **Figure 6e**.

2.5 Filtration

Filtration consists in the separation of the solids contained in a suspension by means of a perforated plate (filtering medium), which allows the passage of the

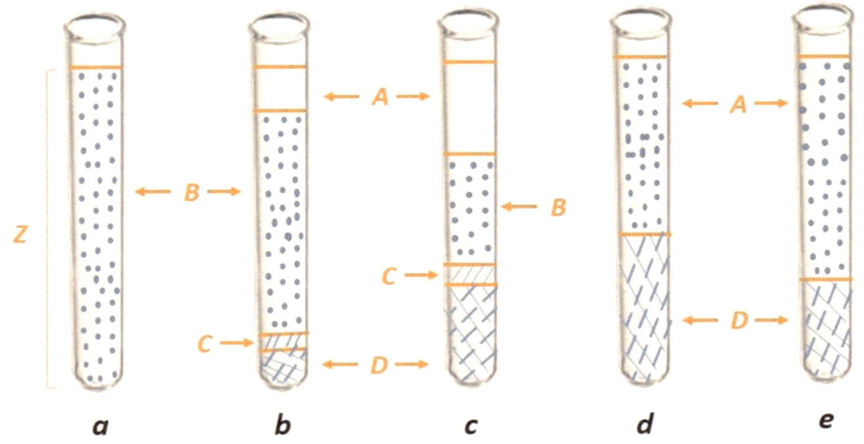

Figure 6.
Evolution of the sedimentation process.

liquid and retains the solid particles. The solid-liquid suspension that is fed to the filter is called syrup; the liquid stream that passes through the filter medium and which is obtained as a product is known as the filtrate. The retained solids form a bed or cake, whose porosity depends on the characteristics of the product to be filtered and the operating conditions, whose thickness increases throughout the filtration process. During the formation of the first layers of the cake, particle bridges are formed which block the perforations of the filter material. In order to overcome the resistance of the cake and the filter medium to the circulation of the liquid phase, a pressure difference must be established between both sides of the filter medium [45].

After the maceration, the previous step was gently filtered with sieve and canvas cloth, manually applying pressure until obtaining liquid rich in pectic substances. The remaining solid is discarded.

2.6 Precipitation with ethanol

The pectin was precipitated from the extract juice by specific means, either precipitation with ethanol or salifying with aluminum chloride [46, 47]. In this work ethanol was added to 96% vol., in a ratio of 70/30 (filtrate/alcohol), to precipitate the pectin. The precipitated pectin was separated from the precipitating solution by filtration. In this case, a canvas cloth filter was used again.

2.7 Drying

The pectin, obtained in the filtrate, was placed in plastic trays and placed in an oven at a constant temperature of 60°C for 48 h to eliminate the excess moisture. Then, they were kept in a sealed container until they were used. The flour final humidity cannot be greater than 5% by weight.

3. Characterization of pectin flour

3.1 Thermal analysis

The generally accepted definition of thermal analysis covers the group of techniques in which a physical property of a substance or a material is measured as a function of temperature while it is subjected to a controlled temperature program. More than a dozen thermal methods that differ in the measured properties and in the temperature programs can be distinguished. The effects of heat on the materials are numerous among which loss of mass, structural changes, conformational changes, changes of state, chemical reaction, degradation, etc. can be mentioned. In thermal analysis, weight changes form the basis of thermogravimetry (TGA), while the measurement of energy changes is the basis of differential thermal analysis (DTA, for its acronym in English) and differential scanning calorimetry (DSC, for its acronym in English). Thus, for example, thermogravimetry gives information about the loss or gain of weight of a sample, while the DTA and the DSC provide information on the variation of heat in a process such as a reaction or physical change, indicating whether it is endothermic or exothermic.

Thermal analysis is the series of techniques where the physical properties of a material are measured as a function of time, while the material is subject to a temperature program. Physical properties are also measured as a function of temperature, when it is variable. In all calorimetric techniques, the measured property is heat.

These methods find broad application in both quality control and research of pharmaceuticals, clays and minerals, metals and alloys, and polymers and plastics [48, 49].

Then, the thermoanalytical methods used in the characterization of the compound of interest of this work were defined and referenced to measurement principles of the instruments occupied.

3.1.1 Differential scanning calorimetry (DSC)

This technique allows the study of those processes in which enthalpic variation takes place, determining the temperatures where physical or chemical changes take place, points of crystallization and boiling, enthalpies of reaction, and determination of other transitions of first and second order [49].

It is a technique in which the difference in the rate of heat flow (or power) to the sample and to the reference sample is monitored against time while the samples are exposed to a temperature program [50].

The instrument is a differential scanning calorimeter, and the main components of any DSC instrument are shown in **Figure** 7.

The purpose of differential scanning calorimetry is to record the difference in the enthalpy change that takes place between the sample and an inert reference material as a function of temperature or time, when both are subjected to a controlled temperature program. The sample and the reference standard are housed in two identical wells that are heated by independent resistances. This makes it possible to use the principle of "zero balance" of temperature. When a thermal transition occurs in the sample (a physical or chemical change that results in a release or absorption of heat), thermal energy is added to either the sample or the reference, in order to maintain both at the same temperature. Because the thermal energy is exactly equivalent in magnitude to the energy absorbed or released in the transition, the energy balance provides a direct calorimetric measurement of the energy of the transition. The energy difference required to maintain the two sample cells at the programmed temperature is the quantity that is represented as a function of

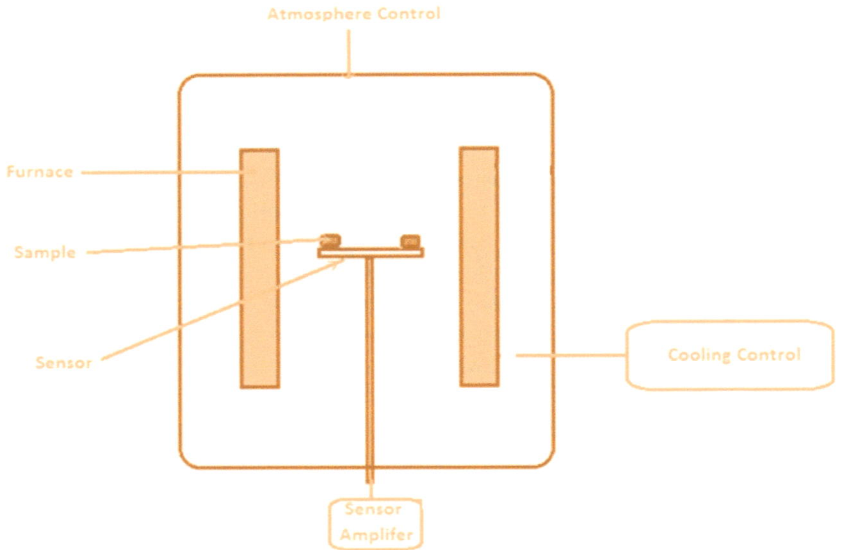

Figure 7.
DSC instrument.

temperature or as a function of time at constant temperature. These two representations are called thermograms [50, 51].

Figure 8 shows the most important events that occur when a synthetic polymer is measured by DSC. These are often characteristic of a substance and serve as a fingerprint, allowing them to be used for quality control [51]. The typical first heating curve of polymers as it shows is the glass transition, cold crystallization, and fusion. The vitreous transition exhibits enthalpy relaxation, which is shown by the overlapping endothermic peak. The latter occurs when the sample has been stored for a long time at a temperature below the glass transition. Cold crystallization occurs when the sample cools rapidly and does not have time to crystallize during the cooling phase. The DSC curve can also be used to determine the specific heat capacity, Cp. A general scheme of a thermogram is shown for a typical semicrystalline polymer, which has been rapidly cooled down to a temperature below its T_g (glassy temperature), the thermogram then being obtained at a certain rate of heating [51].

3.1.1.1 Process in this work

In the differential scanning calorimetry technique, two capsules are available. One of them contains the sample to be analyzed, and the other one is left empty and is the so-called reference capsule. 5 ± 0.5 mg of powdered pectin is weighed, previously ground with glass mortar, and it is placed in one of the aluminum crucibles which is inert. To both capsules an aluminum lid is added, so as not to contaminate the sample. The sweep gas is opened, which in this case is nitrogen with a flow rate of 50 ml/min, and the software is programmed at a sweep temperature with a progressive increase of 5°C/min until it reaches 400°C, while the team records the results for the subsequent analysis.

The measurement was made for each pectin that was obtained and compared to the analysis of a commercial pectin (see **Figure 9**).

According to DSC analysis, the endothermic property of pectin was affected by extraction temperature, while the exothermic property of pectin was only affected by its constituents and raw material [52] (**Table 1**).

The effects of extraction temperature and raw material on the thermodynamic properties of pectin were examined by DSC between 20 and 400°C. As shown in **Figure 9**, an endothermic peak and three exothermic peaks were observed in the DSC thermograms of all pectin samples [52]. The parameters of the peaks were listed in **Table 2**, such as glassy temperature (T_g), melting temperature (T_M),

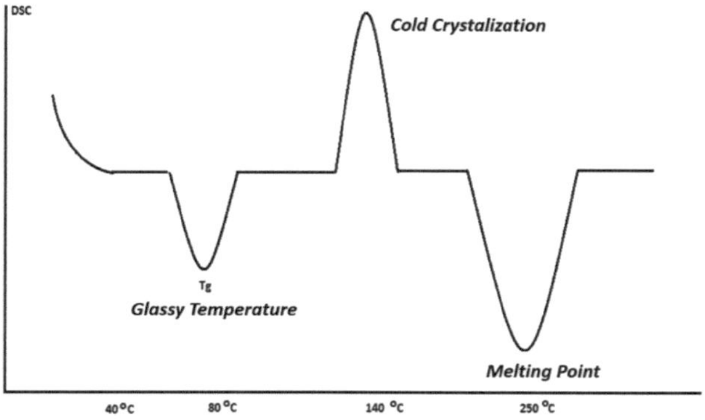

Figure 8.
Main effects measured in DSC using the polyester sample. Temperature range 30–300°C; heating rate 20 K/min; purge nitrogen gas at 50 ml/min.

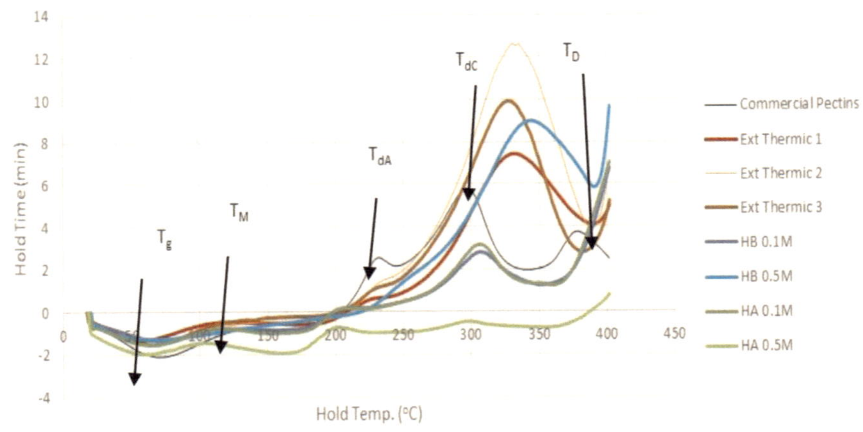

Figure 9.
DSC of pectin produced in this work.

Pectin type	T_g (°C)	T_M (°C)	T_{dA} (°C)	T_{dC} (°C)	T_D (°C)
Commercial pectin	70.81	119.5	234.73	299.63	380.89
HA 0.1 M	62.54	101.3	199.73	309.10	—
HA 0.5 M	58.90	101.8	200.00	300.00	—
HB 0.1 M	69.83	100.7	205.20	308.11	—
HB 0.5 M	69.85	101.6	183.27	342.48	—
Ext. thermic 1	59.74	124.0	223.48	334.17	—
Ext. thermic 2	68.81	133.1	236.45	336.66	—
Ext. thermic 3	68.89	128.7	230.85	332.81	—

Table 1.
Data of DCS of pectin obtained produced in this work.

Group	Band (cm^{-1})	Type	Mode
O-H	3600–3100	Stretching	—
C-H	3000–2800	Stretching	—
Pyranose	1200–950 1149, 1104, 1076, 1052, 1019, and 1016	Resonant absorption energy	Cycle vibrations
COO^-	1617 and 1384	Stretching	Asymmetrical and symmetrical stretching vibrations
COOH	1750	Stretching	Mode non-ionized methylated or protonated carboxyl
COOMe	1750	Stretching	Idem
R-O-R	1200–1100	Absorption ether and pyranosic ring	—
C-C	1200–1100	Stretching	Idem
Galacturonic acid	1120–990		Typical pectin group

Table 2.
FTIR signals of pectin.

deacetylation temperature (T_{dA}), decarboxylation temperature (T_{dC}), and degradation temperature (T_D).

As can be observed in both the figure and the table, the different hydrolyses affect the thermal parameters of the obtained pectin. Analyzing the glass transition temperature data, the T_g of the obtained pectins was less than the T_g of commercial pectin, the T_g for HA 0.5M decreases substantially due to a strong deacetylation of the pectin, on the other hand for the basic hydrolysis it approaches to the value of commercial pectin, this is due to the incorporation of sodium cation as pectinate salt. T_M was very similar for basic and acid hydrolyses but less than the commercial pectin. Thermal hydrolysis resembles the T_M value with respect to commercial pectin. The deacetylation temperature varies from 183.27 to 236.45°C; these values depend on many experimental variables during the hydrolysis, which ultimately results in less or more acetyl groups released. Decarboxylation is another thermal variable intimately related to the previous one and varies from 299.63 to 336.66°C. DSC thermal studies of pectin with other components can be found in the following references [53–58].

3.1.2 Thermogravimetric analysis

TGA is a technique in which the change in the mass of the sample is analyzed while the sample is subject to a change in temperature. The usual thing is that there is a loss of weight; however, it is also possible that there is a gain in weight in some cases. It is controlled in a specific atmosphere. The atmosphere can be static or dynamic with a given flow rate (reduced pressure conditions are also used), and the most common gases are nitrogen, air, argon, and carbon dioxide. Hydrogen, chlorine, and azure dioxide are also used. A fundamental characteristic of the TGA technique is that it only allows to detect processes in which a weight variation occurs such as decompositions, sublimations, reduction, desorption, and absorption while not allowing to study processes such as mergers and phase transitions [49].

As a result of the thermogravimetric analysis, the mass change data are obtained with respect to the temperature or time and a thermogram, which graphically represents the percentage variations of the mass. It should be clarified that this method does not allow to know the chemical composition of the material under study or to identify the thermal changes that are not associated with mass variations such as crystallization or glass transition [50].

Most TGA curves are produced by weight loss, whose main reason is usually [49]:

- Chemical reactions (decomposition and separation of water of crystallization, combustion, reduction of metal oxides)
- Physical transformations (evaporation, vaporization, sublimation, desorption, desiccation)

3.1.2.1 Process in this work

A TG 2950 equipment was used, TA Instruments, Inc., New Castle, USA. The samples are loaded in aluminum cells sealed with a lid of the same material that prevents contamination of the oven in case of dilatation or decomposition of the sample. Each of the pectin obtained was ground with a glass mortar until it reached a fine powder. Here 5 $\pm$ 0.5 mg of sample was used and placed in a platinum crucible, as in the DSC technique (of aluminum or platinum). Then, the equipment was started. The scavenging gas was also nitrogen at 50 ml/min, and previously a purge gas was also used, which was allowed to flow before the start of the experiment. The latter was used to remove gases that may damage the sample and also nitrogen, with a flow of 30 ml/min. As in DSC, the software is programmed at a sweep temperature with a progressive increase of 5°C/min up to 400°C. This experience is also performed for each obtained pectin and commercial pectin, with which it will be compared.

All of the samples presented a characteristic three-step thermal degradation, typical of pectin. In pure pectin, the first step, occurring at about 80°C, corresponds to the water loss; then, it is followed by the second step, between 200 and 400°C (see **Figures 10–12**). In this temperature range, it has been reported that the degradation is primarily derived from pyrolytic decomposition. It consists in a primary and secondary decarboxylation involving the acid side group and a carbon in the ring [59].

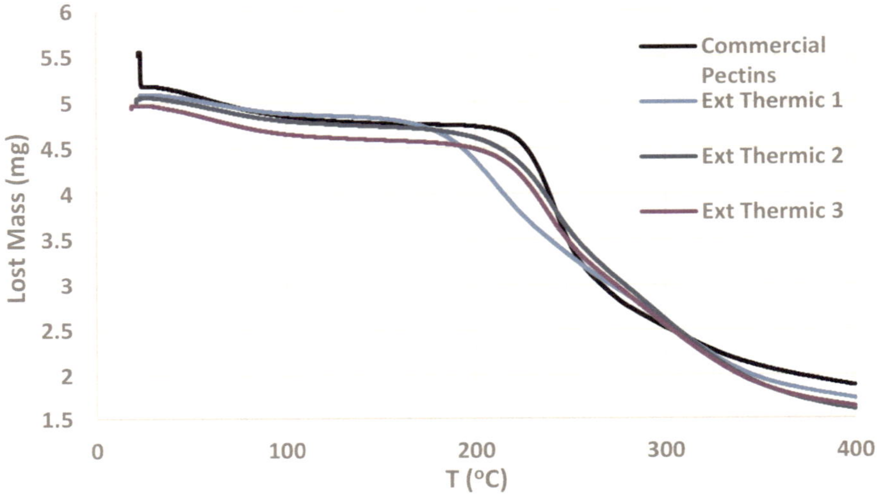

Figure 10.
TGA of commercial pectin and pectin obtained by thermic extraction.

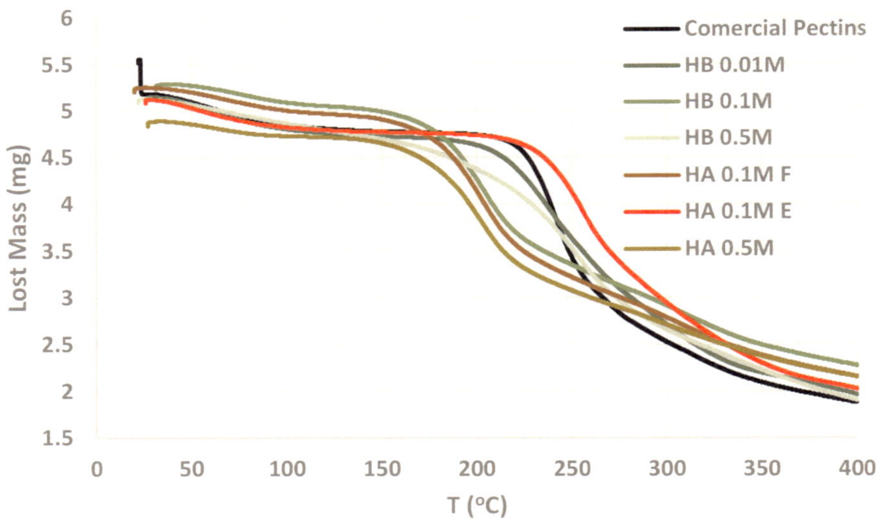

Figure 11.
TGA of commercial pectin and pectin obtained by acid/basic hydrolysis.

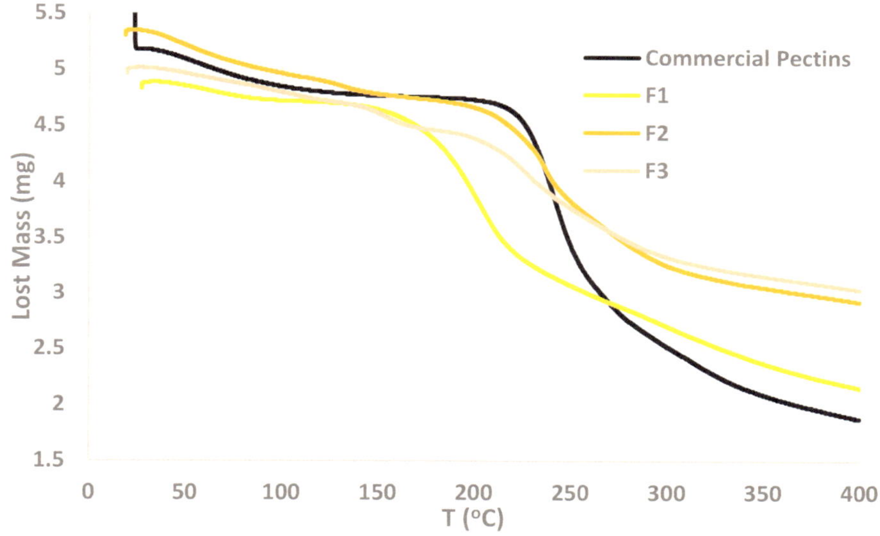

Figure 12.
TGA of commercial pectin and pectin obtained by acid hydrolysis in different fractions.

Moreover, the hydrolysis produces an increase in the evaporation temperature, going from 80°C for commercial pectin to higher values for all the hydrolyzed pectins. The water evaporation step was greatly reduced in these samples, indicating that a very small quantity of water was absorbed. The second degradation stage was very similar for either the pure or the modified pectin samples, although a slightly lower mass loss and a lower midpoint temperature were observed for the modified samples. The third stage, due to oxidative reactions, was postponed, particularly for linoleic residues, as expected for the oxygen scavenging effect of the double bonds [59]. To know more about different thermogravimetric studies of pectins with other compounds, the following references [60–69] can be read.

3.2 Infrared spectroscopy (FTIR)

Infrared spectroscopy is a technique of chemical and structural analysis since it allows the rapid identification of functional groups that are part of an organic and inorganic molecule, being able to perform not only qualitative but also quantitative determinations. This procedure is based on the Michelson interferometer (initially developed to accurately determine the speed of light) and the method of the French mathematician Fourier that allows to convert the obtained information (interferogram) into a spectrum. The infrared region of the electromagnetic spectrum extends between the visible zone and the microwave zone.

The most useful practical section of the extensive IR region is that between 4000 and 650 cm^{-1} called the middle infrared region. The use of the far-infrared (NIR) region, between 650 and 200 cm^{-1}, has expanded considerably in recent decades, especially for the study of organometallic or inorganic compounds (heavy atoms, weak bonds). The near-infrared (NIR) region, between 12,500 and 4000 cm^{-1}, accessible to quartz optics, has been used for quantitative determinations rather than structural purposes. The existence of these characteristic bands of functional groups allows a wide use of IR spectroscopy in the structural determination as well as in the identification of molecular interactions, formation of new bonds, and monitoring of reactions, among others [70, 71].

The most relevant characteristics of this spectroscopy are the following [71]:

1. If two molecules are constituted by different atoms or have different isotopic distributions or configurations or are in different environments, the infrared spectra will be different.

2. A defined substance can be identified by its infrared spectrum. These spectra can be considered as the fingerprints of said substance.

3. The spectra show bands that are typical of particular functional groups and that have specific locations and intensities within the infrared spectra.

4. Molecular structures can be inferred from the spectra. For this a model is required on which to base the calculations.

5. The intensities in the spectrum bands of a mixture are generally proportional to the concentrations of the individual components. Therefore, it is possible to determine the concentration of a substance and perform analysis of samples with several components.

3.2.1 The Fourier-transform infrared spectrometer (FTIR)

A Fourier-transform spectrometer consists of three basic elements: a light source, a Michelson interferometer, and a detector [71].

Its operation is as follows: a collimated beam, coming from a source that emits throughout the infrared region, hits a beam splitter. The incident beam is divided into two perpendicular beams of equal energy, one of which falls on the moving mirror and the other on the fixed mirror. The beams are reflected by both mirrors and recombine upon reaching the beam splitter. This results in an interference, which can be constructive or destructive depending on the relative position of the moving mirror with respect to the fixed mirror. The resulting beam passes through the sample, where a selective absorption of wavelengths occurs and finally reaches the detector. The information collected by the detector is used to obtain the interferogram, which is digitized. A computer develops the approximate calculation of the Fourier transform of the interferogram [71].

Solid materials generally exhibit too much absorption to allow direct transmission of infrared radiation. Only in some cases can very thin films of the material be obtained that allow, without mixing it with others, to obtain a spectrum by transmission. In solids in the form of dust, in addition to the problem of absorption, another occurs: much of the radiation transmitted is scattered. Since the dispersion is proportional to the difference between the refractive indices, some improvement is obtained by placing the finely pulverized solid in a liquid medium whose refractive index coincides with that of the substance. The medium serves at the same time as a diluent. Often hydrocarbon oil (nujol), or oil from a fluorocarbon polymer (fluorolube), is used for this purpose; both have their own absorption bands. To prepare this sample sprayed in oil, milligrams of the powder are ground in a drop of oil until a very fine paste is obtained, which then spreads like a thin film between two layers of sodium chloride (NaCl). Another technique for preparing diluted solid samples is the potassium bromide tablet (KBr) method. The sample is milled with powdered potassium bromide and then compressed in a die, at a pressure of 700 kg cm^{-2} ($\sim 7 \times 10^7$ Pa), in a hydraulic press. The thin disk thus formed is sufficiently transparent and only shows appreciable dispersion at wavelengths less than 10 μm. Since potassium bromide is hygroscopic, and it is almost impossible to eliminate atmospheric water, in the samples prepared by this method, the OH band is always observed. Other materials than potassium bromide can be used; sometimes polyethylene powder offers advantages [71].

3.2.1.1 Process

The technique used was the preparation of the sample with potassium bromide. A pinch of pectin, primarily ground in a glass mortar, was mixed with potassium bromide. The resulting mixture was compressed through a hydraulic press, and a thin pellet was formed, which is placed in the spectrometer to perform the corresponding analyses.

FTIR spectra provided structural information of pectin in the region between 4000 and 400 cm^{-1}, where the major chemical groups in the pectin were identified (see **Figures 13–15** and **Table 2**).

For the pyranose cycle vibration region, one should note almost identical spectral parts with bands at 1149, 1104, 1076, 1052, 1019, and 1016 cm^{-1} characteristic for peptic substances. The band at 1750 cm^{-1} is assigned to stretching C=O mode non-ionized methylated or protonated carboxyl. Ionization, i.e., the formation of

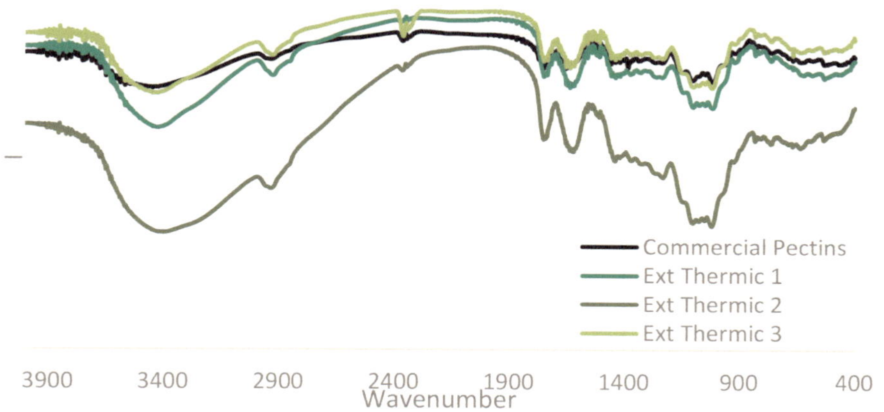

Figure 13.
FTIR of pectin thermic extractions.

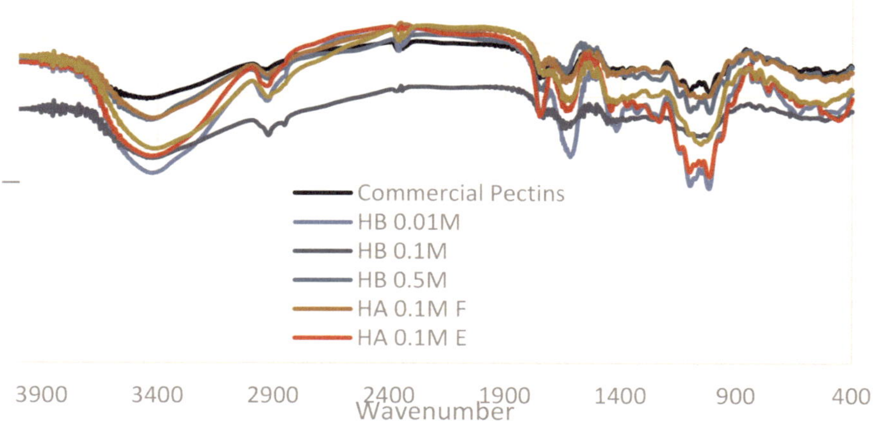

Figure 14.
FTIR of pectin obtained by acid/basic hydrolysis extractions.

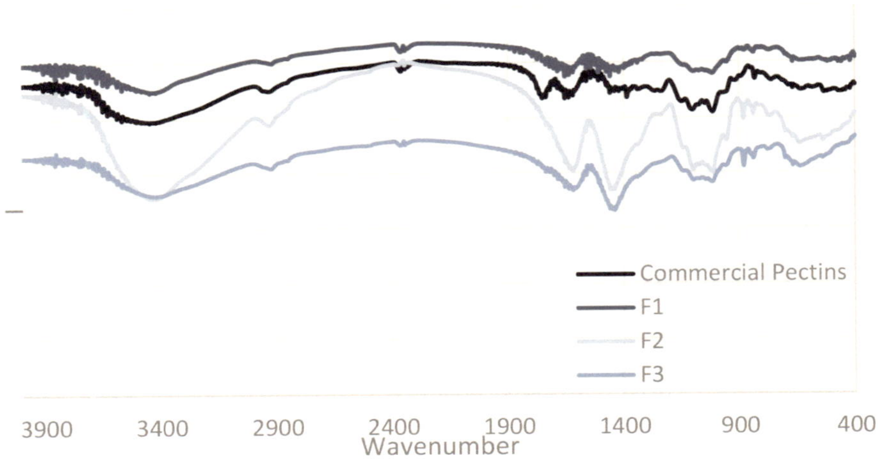

Figure 15.
FTIR of pectin fractions.

salts, leads to its disappearance, and two new bands appear due to asymmetric and symmetric stretching modes of COO^- at 1650–1600 and 1450–1400 cm^{-1}, respectively. The absorption bands between 1200 and 1100 cm^{-1} were from ether (R-O-R) and ring C–C bonds in pectin molecules. The selection of the most important wave numbers, by two independent chemometric techniques, allowed to define the region between 1120 and 990 cm^{-1} as the range for the spectral identification of galacturonic acid in peptic polysaccharides. There are two bands in the quince pectin spectrum within this region, a major one centered at 1617 cm^{-1} and a less intense one at 1384 cm^{-1}. These two bands correspond, respectively, to asymmetrical and symmetrical stretching vibrations due to the COO^- group of polygalacturonic acid. The absorbances at 1104 and 1000 cm^{-1} are the galacturonic acid, because all peptic polysaccharides are characterized mainly by these peaks (see **Table 2**) [72–76].

In general traits, hydrolyzed pectins have a greater degree of affinity for water due to the -OH groups that increase greatly; this may also be due to the fact that the flours were not dried correctly, but even so a slight and positive tendency can be observed. Another positive aspect is that the hydrolyzed pectins have an increase in the C=O bands because they possess a more marked degree of acetylation, although it cannot be calculated analytically. The typical galacturonic acid signal in 1618, 1388, and between 1000 and 1105 cm^{-1} is observed and specifically characterizes the pectin. On the other hand, there are no substantial differences between commercial pectin and the different hydrolyzates obtained, neither between these and other pectins reported in the literature [77–79] (see **Figures 13–15**).

4. Characterization of pectin solutions

4.1 Densimetry

The measurement of the density is a complementary measure to the viscosity and allows to calculate the partial specific volume and together with the viscosimetry allows to determine the hydration value of the macromolecule in solution.

The density data of the hydrolyzed pectin solutions have a $\partial\rho/\partial c$ ratio similar to that of pectin commercial but with different order at the origin. This phenomenon

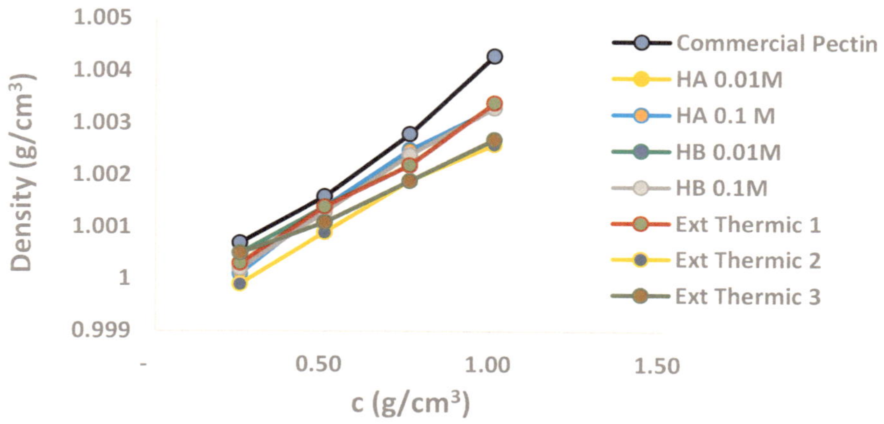

Figure 16.
Density data of pectin solutions.

is attributed to the fact that the commercial pectin has a molecular weight superior to that of the different hydrolysates, which ultimately translates into a greater capacity to incorporate water in its polymeric structure, that is to say it is more hydrophilic than the rest (see **Figure 16**).

4.2 Viscosimetry

Viscosity is one of the most important properties of polymer solutions. The viscosity depends on the chemical structure of the polymer, the interactions with the solvent, and the molecular weight. Normally, a molecule of high molecular weight in a good solvent acquires a large hydrodynamic volume, and the viscosity of the solution increases [80].

The viscosimetry of diluted solutions is related to the measurement of the intrinsic ability of a polymer to increase the viscosity of a solvent at a given temperature and is useful for obtaining information related to the size and shape of the polymer molecules in solution and the polymer-solvent interactions. In the diluted regime, the viscosity of a polymer solution (for very low polymer concentrations) is determined relative to the viscosity of the solvent. The following terms are defined in these cases:

$$\text{Relative viscosity, } \eta_r = \frac{\eta}{\eta_0} \tag{1}$$

where η is the viscosity of the polymer solution and ηo is the viscosity of the pure solvent.

$$\text{Specific viscosity, } \eta_{sp} = \eta_r - 1 = \frac{\eta - \eta_0}{\eta_0} \tag{2}$$

$$\text{Reduced viscosity, } \eta_{red} = \frac{\eta_{sp}}{C} \tag{3}$$

where C is the concentration of the polymer. This is a measure of the ability of the polymer to increase the viscosity of a solvent.

$$\text{Inherent viscosity, } \eta_{inh} = \frac{\ln \eta_{red}}{C} \tag{4}$$

Even in highly diluted solutions, polymer molecules are capable of forming intermolecular interactions. The two contributions to reduced viscosity are the movement of the isolated molecules in the solvent and the interaction between the polymer molecules in the solution. To eliminate the interactions, it is necessary to extrapolate to zero concentration to obtain the inherent and reduced viscosities commonly known as intrinsic viscosity:

$$[\eta] = (\eta_{red})_{c\to 0} = (\eta_{inh})_{c\to 0} \tag{5}$$

The intrinsic viscosity has the units of mass / volume, it is a measure of the size of a molecule in solution, based on the ability of a polymer molecule to increase the viscosity of a solvent in the absence of intermolecular interactions [80].

The most common equations for evaluating the intrinsic viscosity are the Huggins and Kraemer equations, given by Eqs. (6) and (7), respectively. The most usual procedure for determining the intrinsic viscosity is to determine the relative viscosity for different concentrations of polymer and to represent the data using Eqs. (4) and (5) and then calculate the value at zero concentration. **Figure 15** shows a typical plot of this type of data [80]:

$$\frac{\eta_{sp}}{c} = [\eta]_H + k_H\,[\eta]_H{}^2 c \tag{6}$$

$$\frac{\ln \eta_r}{c} = [\eta]_k + k_k\,[\eta]_k{}^2 c \tag{7}$$

Mark and Houwink, M-H, independently correlated the intrinsic viscosity with molecular weight, and this equation is applicable to many polymers and biopolymers and is used to determine molecular weight. The *k* and *a* parameters both vary with the nature of the polymer, temperature, and solvents. The calculation of M-H parameters is carried out by the plot representation of the following equation:

$$\ln[\eta] = lnk + alnMW_v \tag{8}$$

The exponent *a* is a function of polymer geometry and varies from 0.5 to 2.0. These constants can be determined experimentally by measuring the intrinsic viscosity of several polymer samples for which the molecular weight has been determined by an independent method (e.g., diffusion, osmotic pressure, sedimentation equilibrium, and light scattering).

4.2.1 Process

A capillary viscometer Ubbelohde IVA 1C of glass was used (**Figure 17**) where the time required by the two fluids to run between two marks in a capillary was recorded. Therefore, a digital chronometer was also required to take the runoff time, and an Anton Paar DMA 35 M density meter was used to determine the density of water and solutions.

Diluted solutions of the different pectins were prepared with distilled water at 0.25, 0.50, 0.75, and 1% wt. Each of them was stirred for a few minutes, under temperature conditions (50°C), to form a homogeneous solution. And before starting the measurements, it was allowed to cool to approximately room temperature.

The runoff time of each solution was taken by loading the viscometer, starting with the most diluted and continuing with the most concentrated. Previously the runoff time of the water that was used was taken. All measurements were made in triplicate to calculate an average. In addition to the times, density and temperature

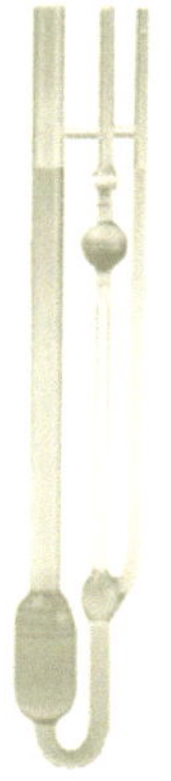

Figure 17.
Ubbelohde capillary viscometer.

are measured for each solution and for the solvent, to then perform the corresponding calculations, taking into account these variables.

The intrinsic viscosity is obtained at the intersection of the line with the axis of the ordinates, that is, when the concentration tends to zero.

The pectin commercial has an intrinsic viscosity of 284.14 cm^3/g, which is lower than other data reported in previous studies [81, 82]; this is due to the fact that there are different hydrolyses and pectin purification treatments.

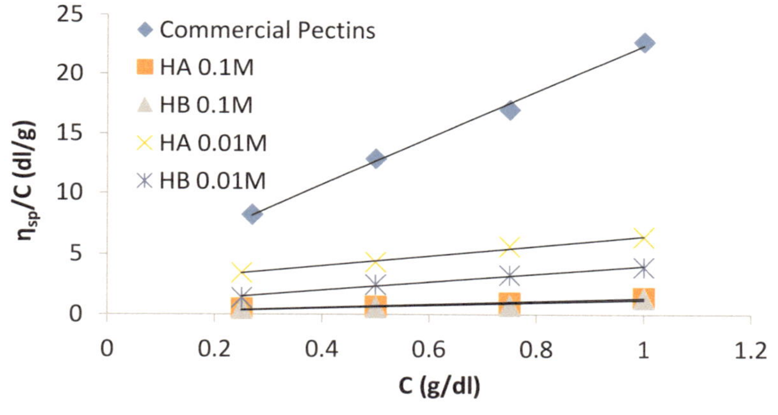

Figure 18.
Evaluation of the intrinsic viscosity for pectin solutions at 25°C.

Hydrolysis	c (mol/L)	[n] (cm^3/g)	k_H	M (g/mol)	R_H (nm)	$\tilde{v}$	δ (g/g)	h
Commercial pectin		284.14	19.69	92928.83	12.12	0.51	26.48	3.75
Acid	0.01	237.93	4.10	74883.61	10.63	0.57	22.03	3.41
	0.1	144.88	3.01	40956.73	7.37	0.64	13.12	2.78
	0.5	7.79	1.25	1169.89	1.85	0.73	0.03	1.01
Basic	0.01	187.34	2.05	55988.91	8.91	0.58	17.22	3.13
	0.1	62.86	3.39	14832.63	3.98	0.51	5.46	2.27
	0.5	7.25	1.11	1072.00	1.25	0.49	0.20	1.12

Table 3.
Hydrodynamic data of hydrolyzed pectins (for data, read references [81, 82]).

As for the different hydrolysates, both basic and acid, the intrinsic viscosity data are lower than the data of commercial pectin, and in turn these decrease with the increase of acid or base added to the hydrolysis. The hydrodynamic radius (R_H) and the hydration value (δ) decrease for the same reason in addition to the fact that the molecular weight also decreases. Another aspect not less is that for the same hydrolysis, but different reagent, be it hydrochloric acid or sodium hydroxide, the salt formed in the latter is more soluble in water. A consistent phenomenon is the increase of the partial specific volume ($\tilde{v}$) and is reasonable due to the increase of the hydration value. Finally, it can be said that K_H values show that water is a good solvent for pectins (see **Figure 18** and **Table 3**).

5. Film synthesis and characterization

5.1 Film synthesis

The technique used for the preparation of the membranes is the "casting" method. This consists of depositing a polymer solution on a mold or support of a given material, allowing it to evaporate at room temperature, or under some air flow or N_2, to obtain a membrane with a certain thickness and a surface area. The polymer is first dissolved in a solvent to form the solution, and then some other compounds, such as plasticizers, antioxidants, stabilizers, etc., are added to the polymer solution and dispersed by agitation. The size of the container and the polymer concentration will determine the thickness and length of the membrane; it is necessary that in the molding there are no bubbles and no bubbles are generated during the solvent evaporation process. Another important operative aspect for this procedure is the evaporation rate of the solvent and a strict control of the temperature. Subsequently, the membrane is molded by removing the solvent.

This method is the most used in the preparation of membranes at laboratory scale in order to study the influence of the composition and morphology of the materials on the behavior of the films. This technique has a great operational ease and is suitable for all types of soluble materials in a particular solvent. The concentrations of polymer and other added components can also be easily controlled.

5.1.1 Technique

Once the pectin flour was obtained in the previous step, a 2% by weight aqueous solution was prepared with 0.5% by volume of glycerin, stirring until total dissolution. The solution is then placed in a flat mold 10 cm in diameter and put into an oven at a temperature of 50°C for 24 h. Slow evaporation of the solvent is preferable in order to obtain a dense membrane. It is then demolded and placed between two glass plates for 48 h to avoid deformation of the surface of the membranes. In this last period, the solvent is finished evaporating.

Because the packages have multiple functions to which they are used to pack a great diversity of foods, their characteristics and properties are very varied. The materials used to make them are as different as paper, cardboard, plastics, glass, tin, aluminum, and combinations thereof. As for its shape, they are found as boxes, cans, bags, bottles, and movies, among others. The latter are defined as continuous, thin matrices that are structured around the food they protect. Most commercially used films are made with different types of plastics, which, besides being obtained from hydrocarbons, have the disadvantage of not being biodegradable [83–86].

A biodegradable packaging is defined by the ASTM as one that is capable of decomposing into carbon dioxide, methane, water, inorganic compounds, or biomass, the dominant mechanism of decomposition being the enzymatic action of the microorganisms, and the resulting products can be obtained and measured in a given period of time [87].

The materials used for the production of biodegradable packaging can be polymers of natural origin (proteins, starch, lipids, and chitosan, among others) or of synthetic origin (i.e., polyhydroxyalkanoates and polylactic acid). Among the polymers of natural origin that are being used for the production of biodegradable films are pectins [87].

Plasticizers are normally used to facilitate the processing and/or increase the flexibility of the film. Water, some oligosaccharides, polyols, and lipids are different types of plasticizers widely used in films based on hydrocolloids. Their combination could lead to synergistic effects between the components improving the properties of the films. Various theories have been raised about how these compounds work. The theory of lubrication postulates that plasticizers intermix and act as internal lubricants by reducing the frictional forces between the polymer chains. The theory of the gel postulates that the rigidity of the polymeric network comes from its three-dimensional structure; then the plasticizers would act by breaking polymer-polymer interactions that are interposed between them. The theory of free volume postulates that the addition of plasticizers is a way to increase the free volume, reducing the interactions between the chains. All the proposed theories agree that the effects generated by the addition of glycerol are due to these small molecules that are located between the polymer chains [87].

The technique used for the formation of the films was the following: A sample of 1 g was mixed with 50 ml of distilled water and 0.5 ml of glycerin. It was stirred with mechanical agitator for half an hour at 40°C. The identified petri dish solutions were overturned, so that only the surface of the petri dish was covered. It was taken to the stove at 80°C for 48 h. The formed films were demolded and kept in sealed envelopes. After obtaining the films, tests were conducted to evaluate their properties against a commercial pectin.

5.2 Scanning electron microscopy (SEM)

This technique is used to inspect in a relatively easy way the morphological features of the membranes, especially since this technique is very suitable to evaluate if the film is dense or porous and the pore size distribution and elemental analysis (EDS) are a standard procedure to identify and quantify the elemental composition of sample areas of up to several cubic micrometers. The sample material is bombarded with electrons in a scanning electron microscope, and the X-rays produced are measured with an X-ray spectroscope. Each element has a characteristic wavelength through which it can be identified.

The images of an electron microscope are obtained by the detection, processing, and visualization of the signals resulting from the interactions between a high-energy electron beam and matter. These interactions can provide information on topography, composition, and crystallographic structure [87].

In a scanning electron microscope, the image is obtained from the signals emitted by the sample, and it is formed as the electron beam moves on a portion of its surface. This scan is performed line by line on a small area of rectangular shape (raster). This zone is the one that is visualized and amplified in the final image. As the beam explores the sample in this way, the intensity of the generated signal varies according to the particular point analyzed at each instant. You can thus obtain images of all types of structural materials or biological material with a minimum of previous preparation and directly observe all types of surfaces with a great depth of focus. This is a unique feature of electronic instruments that allows obtaining micrographs in focus of irregular surfaces as a fracture surface. For this it is only necessary to ensure that the samples are clean, dry, resistant to high vacuum of the instrument, and good electrical conductors. If it is a question of observing a non-conductive material, the samples are usually coated with a thin metallic layer, as gold, or the samples that have high water vapor contents are previously dried [87].

The composition of the sample affects both the depth and the shape of the interaction volume. Denser samples (composed of heavy elements) tend to reduce the penetration of the beam and also reduce the distance that the signals generated can pass without being reabsorbed by the sample. The volume of interaction tends to be more squashed and similar to a hemisphere. On the contrary, in less dense samples, composed of light elements, the volume takes on the characteristic shape of a drop [87].

5.2.1 Equipment

All scanning electron microscopes consist of an electronic cannon in a high vacuum column, in the order of 10^{-5} mmHg, in which a high-energy electron beam (5–30 kV) is generated. This beam is collimated by a series of electronic lenses and focused on the sample analyzed. The detectors register the signals originated by the interaction between the electron beam and the sample, which are processed and visualized in the final observation system (monitor or computer screen). The electron gun is the first component of the microscope column and is the one that produces the electron beam. It consists of filament emitting electrons that are then accelerated by an anode positively polarized to a variable voltage between about 5 and 30 kV [87].

Interpretation of the images: the contrast of a micrograph in the secondary electron mode (emissive mode) comes from the variations in the topography of the sample. In effect, the voltage applied between the grid of the detector and the sample favors the collection of secondary electrons on sharp edges since the electric field is more intense there. More secondary electrons can be collected from a projection or step of the sample than from a depression or cavity. The projections then appear brighter than the depressions, which makes the interpretation of the micrographs immediate [87]. The equipment used to obtain the SEM images was a scanning electron microscope JOEL 1450VP.

The superficial images of the selected pectins (**Figures 19** and **20**) show that the commercial pectin (**Figure 21**) is more homogeneous with a morphology similar to

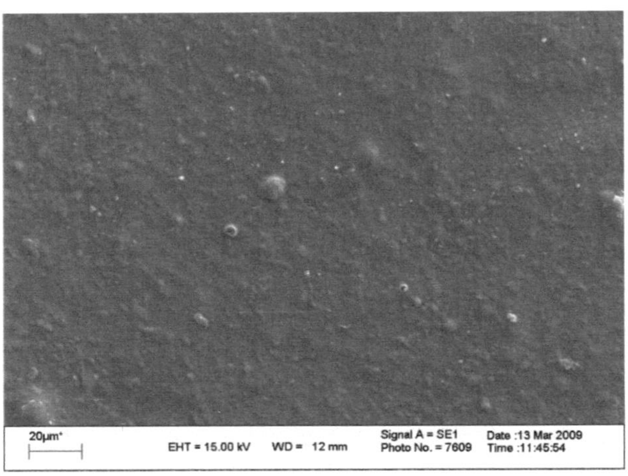

Figure 19.
SEM surface image 500× of HA 0.1 M membrane.

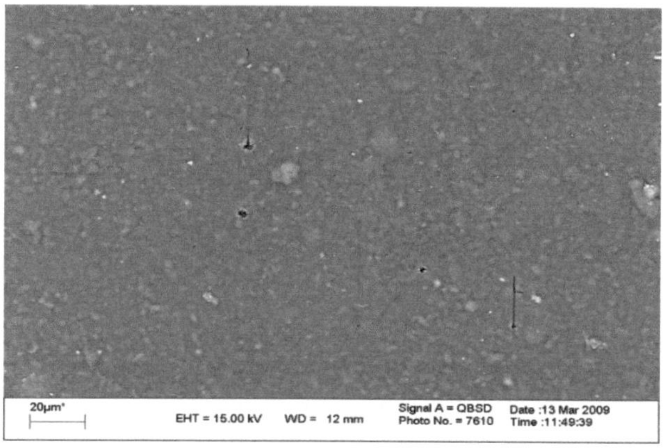

Figure 20.
SEM surface image 500× of HB 0.1 M membrane.

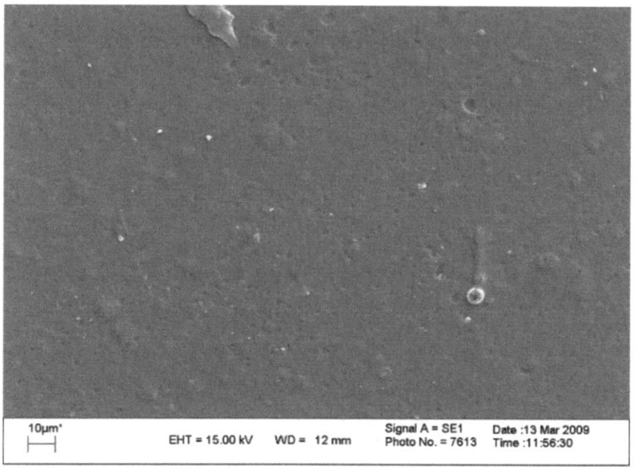

Figure 21.
SEM surface image 500× of commercial pectin membrane.

craters, valleys, and peaks (not acute). On the other hand, both acidic and basic hydrolysates are heterogeneous, where it should be noted that the pectin that the basic hydrolysis film has more rigidity which causes breakage on its surface due to the formation of the sodium pectinate salt.

5.3 Permeation

Unlike glass or metal packaging materials, packages made of plastic are permeable in different degrees to small molecules such as gases, water vapor and organic vapor, and other low molecular weight compounds such as flavorings and additives present in food. As a consequence of the barrier properties of the material, the transfer of this molecule varies from high to low. The knowledge of the behavior solution/diffusion/permeation of these molecules through the polymer film has become more and more important in recent years, especially for polymers used in the field of food packaging where external environmental contamination should be avoided and the shelf life of food controlled by the use of modified atmosphere packaging techniques (MAP). Many factors that can influence the performance of polymer packages must be taken into consideration to design the correct package market solution [88].

Since contact with food can alter the performance of the polymer, it is important to study the characteristics of the material barrier under realistic conditions. For example, the absorption of vapor or ambient liquid can cause increased plasticization of the polymer, resulting in a decrease in mechanical properties. The purpose of this work is to give an overview of the state of the art of the permeability behavior of polymer packages used for food applications [88].

5.3.1 Theory of permeation

The diffusion of the permeate through a film is influenced by the structure of the film, the permeability of the film to specific gases or vapor, thickness, area, temperature, pressure difference, or concentration gradient through of the movie. Permeability, as reported in the literature, is defined as the quantification of permeate, gas, or vapor transmission, through a resistant material. Then, the concept of permeability is usually associated with the quantitative evaluation of the barrier properties of a plastic material. In a flawless material such as pores or cracks, the primary mechanism for the flow of gas and water vapor through a film or coating is an activated diffusion. This means that the permeate dissolves in the matrix of the film in the higher concentration part; diffuses through the film, driven by a concentration gradient; and evaporates from the other side of the surface. Differences in the solubility of specific gases can influence the diffusivity of the gases through the film. The second step, diffusion, depends on the size, shape and polarity of the penetrating molecule, and on the crystallinity, degree of cross-linking and segmental movement of the polymeric chain of the polymeric matrix. [88].

As for the theory, the penetration of gas through a polymer is described by a diffusion model, using the laws of Henry and Fick to obtain the expression that relates the rate of permeability with the area and the thickness of the film [88].

The mechanism can be described in a very simple way as in **Figure 22**, for a homogeneous polymer film with thickness l and permeable pressure p (with $p_1 > p_2$) and as the different concentration permeable through the film (with $c_1 > c_2$).

The flow of permeate (gas or vapor), indicated by J, is described by Fick's first law:

$$J = -D\,\Delta C \tag{9}$$

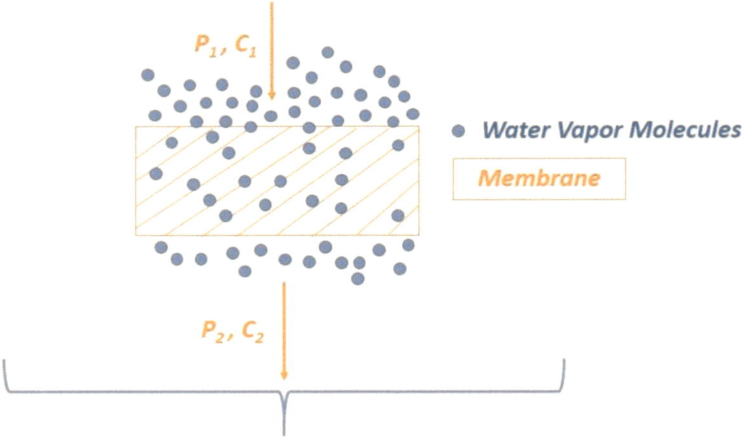

Figure 22.
General mechanism for water vapor permeation through a plastic film.

that, for the one-dimensional diffusion through a membrane polymer and in stationary conditions, it can be written as

$$J = -D\,\Delta C/l \tag{10}$$

where J is the diffusion flow (expressed in mol.cm^{-2} s^{-1}), D the diffusion or diffusivity coefficient (expressed in cm^2 s^{-1}), and ΔC the difference in concentration (expressed in mol cm^{-3}) through the thickness of the membrane l (expressed in cm). D reflects the rate at which the permeant diffuses through the polymer. When the mechanism of diffusion is in its stable state, the equilibrium of the concentration of gas c on the surface and the partial pressure of gas p obeys Henry's law. When the permeant is a gas, it is more convenient to measure the vapor pressure p (expressed in atm); then ΔC can be replaced by $S\,\Delta p$, where S (expressed in mol cm^{-3} atm^{-1}) is the solubility coefficient that reflects the amount of permeant in the polymer and Δp is the pressure difference in the entire film. Eq. (10) becomes

$$J = -D\,S\,\Delta p/l \tag{11}$$

The $D\,S$ product is indicated as a coefficient of permeability (or constant) or permeation coefficient or simply as permeability (PAG).

If S is independent of the concentration, that means a linear relationship between the concentration-distance through the polymer, the coefficient of permeability P can be defined as

$$P = -J\frac{l}{\Delta p} = -D\,S \tag{12}$$

The P/l ratio, indicated by q, is called permeance. The diffusion takes place only in one direction, through the film and not along or through it; in addition, the coefficients D and S are independent of the concentration of the permeate. This behavior of molecular diffusion in the polymer is indicated as Fick behavior. Obviously, as also reported by Robertson, there are many realistic cases where these hypotheses are not valid as, for example, when the steady state takes a long time to reach (as a glassy polymer) or when the coefficients D and S are correlated to the interaction between permeate and polymers, such as the interaction between water and hydrophilic film or similar solvent vapor that diffuses through polymer films. These cases are indicated as a non-Fick behavior [88].

5.3.2 Water vapor permeability

The concept of permeability is normally associated with the quantitative evaluation of the barrier properties of the material. If it does not have defects such as pores or cracks, the primary mechanism for the flow of gases or water vapor through the film is an activated diffusion. This means that the molecules enter the film matrix on the higher concentration side, diffuse through the film following the concentration gradient, and detach from the surface in the low concentration area. The second step of the phenomenon of permeability, diffusion, depends on the size, shape, and polarity of the penetrating molecule; it also depends on the crystallinity, degree of cross-linking, and movement of the polymer chains in the matrix of the film. The gaseous molecules are unable to permeate through the polymeric crystals since they are insoluble in them. Thus, gas permeation in semicrystalline polymers is confined to amorphous regions. The decrease in permeability in materials with a certain degree of crystallinity is due, then, to the smaller volume available for gas penetration and to the long and tortuous trajectory that molecules must make between the crystals. The reduction in the permeability value is proportional to the volume fraction of the crystalline phase.

The permeation of a gas or vapor through the polymer is described by a diffusion model, using the Henry and Fick laws to obtain the expression representing the permeation rate taking into account the area and thickness of the film.

The mechanism can be described for a homogeneous polymer film of thickness λ, permeant pressure p (with $p_1 > p_2$), and permeant concentrations c through the film (with $c_1 > c_2$):

$$\tau = \frac{Q}{A * t} [=] \frac{ng}{m^2\, s} \tag{13}$$

$$P = \frac{\tau.\lambda}{\Delta P} [=] \frac{ng\ m}{m^2\, s\ Pa} \tag{14}$$

where τ is the transmission speed (ng/m^2s); Q, the permeating mass (ng); λ, the thickness of the film (m); A, the area of the cell (m^2); t, the measurement time (s); and ΔP, the pure water vapor pressure, in the case of calculating water vapor permeability (4238.605 Pa), or high pressure in the case of calculating gas permeation.

The pectin hydrolyzed with acid or base shows a lower permeability than the commercial pectin; the exception to the rule is the pectin extracted thermally (see **Figure 23**).

5.3.3 Water sorption

The properties of barrier to the steam of water for the packaging of alimentary products, in which the physical and chemical deterioration is related to his content of humidity in balance, are of big importance to maintain or extend his useful life. In fresh products, for example, it is important to avoid dehydration.

The ASTM E96 standard defines the water vapor permeability as the rate of water vapor transmission through a unit area of a flat material and per unit

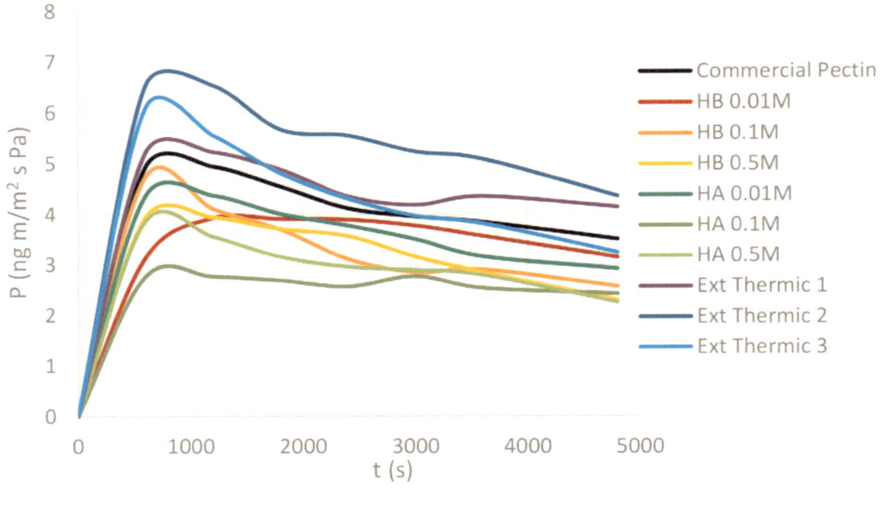

Figure 23.
Water vapor permeation of pectin membranes.

thickness induced by a unitary vapor pressure difference between the two surfaces of the material under study, under certain temperature and humidity conditions.

$$\%SW = 100\frac{w_{sw} - w_M}{w_M} \tag{15}$$

where %SW is the water sorbed percentage (% g/g), w_{sw} is the mass of water (g), and w_M is the mass of the membrane (g).

5.3.3.1 Instrument

The equipment has a hermetically closed chamber that inside has an automatic humidity and temperature controller, both variables controlled by sensors. The humidity range that this sensor can detect is from 0 to 100%, and the accuracy in the measurement is 2% when the relative humidity goes from 0 to 90% and 3% above 90% at 20°C. The temperature and humidity are regulated by turning on or off the controllers of both variables: a focus and a humidifier. In addition, a RADWAG balance with precision of 0.1 mg and a Kolve thickness meter with precision of 0.1 μm were used.

5.3.3.2 Process

From the formed films, samples of 25 mm in diameter were cut, their thickness was taken, and they were placed in the perforated caps of bottles with 20 g of silica gel inside each one. Each bottle is identified by a number that corresponds to a particular pectin. Before starting the experiment, each bottle with silica gel was weighed and the lid placed. Then, they enter the equipment, which must be found at 30°C and 85% humidity, and every 10 min the weights of each one are taken, for 1 h. Then, they are weighed every 20 min. until the cell maintains a constant weight.

In a first stage (see **Figure 24**), which corresponds to a first-order kinetics for water sorption, all the membranes comply, but in the zone of zero-order plateau, they behave in a different way keeping the order. The exception to the rule is the HB 0.5M membrane that corresponds to a pseudo first order and in the plateau area the absorption of water keep increasing, this phenomenon is due to the excess of sodium cations in the pectin (as sodium pectinate) that increase the adsorption of water, and this happens in several stages. The same phenomenon can be observed in the rest of the basic hydrolysis but to a lesser degree.

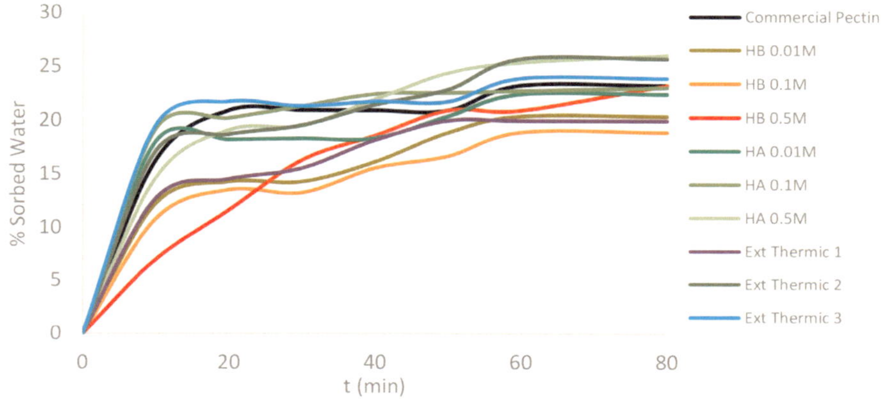

Figure 24.
Water sorption of pectin membranes.

5.4 Mechanical properties

When selecting a polymer to be used as a film, certain characteristics related to its mechanical properties must be considered to ensure that the material successfully fulfills its function. Some of them are tensile strength, elongation, impact resistance, sealing on the seals, resistance to tearing, easy opening of the container, etc.

Materials with the same chemical composition and other similar properties can have very different mechanical properties, depending on their microstructure. In addition, changes in temperature, the cyclical nature of the applied stresses, chemical changes caused by oxidation, corrosion or erosion, microstructural changes due to temperature, the effect of possible defects introduced during machining operations, or other factors can also have a great effect on the mechanical behavior of the materials.

The mechanical properties of a material describe the way in which it responds to the application of a force or load. The force, which acts on the unitary area, generates a deformation. In many materials, elastic stress and deformation follow a linear law. The slope in the linear portion of the strain versus strain curve defines the Young's modulus or elastic modulus of a material. The modulus of elasticity or Young's modulus (E) is the slope of the stress-strain curve in the elastic region (linear portion of the curve) and informs us about the fragility and ductility of the films made.

A material is selected by adapting its mechanical properties to the service conditions required for the component. The first step in the selection process requires that the application be analyzed, in order to determine the most important characteristics that the material must possess. Once the required properties are known, the appropriate material can be selected, using the information contained in the manuals [89]. Then, several tests that are used to measure the way in which a material resists an applied force will be studied. The results of these tests will be the mechanical properties of said material [89].

Tension test: the use of the stress-strain diagram is to measure the resistance of a material to a static or gradually applied force. A test device appears in **Figure 25**; a typical specimen has a diameter d and a calibrated length, l [89].

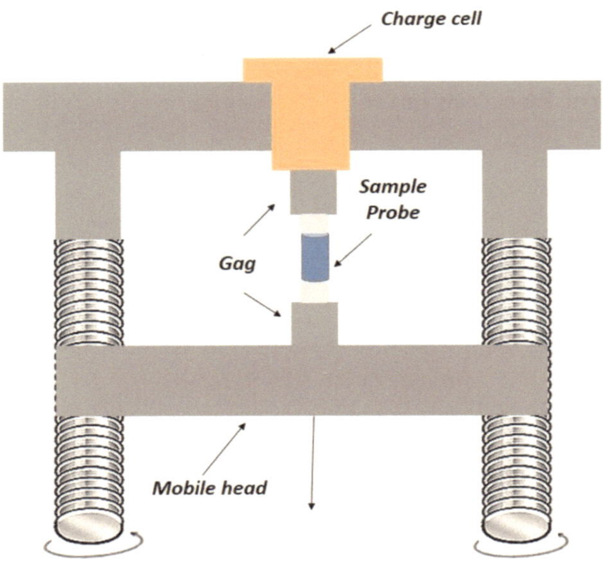

Figure 25.
Tensiometer model used in stress tests.

The test piece is placed between jaws and a force F is applied, which is known as load. To measure the elongation of the material caused by the application of force in the calibrated length, an extensometer is used [89] (see **Figure 25**).

1. Engineering effort and deformation: for a given material, the results of a single test are applicable to all sample sizes and shapes, if the force is converted into effort and the distance between calibrated marks in deformation. The stress σ and the engineering deformation ε are described by the following equations:

$$\sigma = \frac{F}{A} \tag{16}$$

$$\varepsilon = \frac{(l - l_0)}{l_0} \tag{17}$$

where A is the original cross-sectional area of the specimen before the start of the test, l_o is the original distance between calibrated marks, and L is the distance between them, after the force F has been applied [89].

2. Effort of yield: the effort or yield limit is the effort that divides the elastic and plastic behavior of the material. If you want to design a component that does not deform plastically, you must select a material with a high yield strength or make the component of sufficient size so that the applied force produces an effort that is below the yield strength.

3. Resistance to tension: the effort obtained from the highest force applied is the resistance to tension or maximum tension, which is the maximum stress on the engineering strain-deformation curve. In many ductile materials, the deformation does not remain uniform. At a certain moment, one region is deformed more than others, and a significant local reduction occurs in the straight section. This locally deformed region is known as a zone of stricture. Since the area of the cross section at this point becomes smaller, a smaller force is required to continue its deformation, and the engineering force, calculated from the original area A, is reduced. Tension resistance is the effort at which this stricture begins in ductile materials.

Elastic properties: the modulus of elasticity or Young's modulus, E, is the slope of the stress-strain curve in its elastic region. This relationship is Hooke's law:

$$E = \frac{\sigma}{\varepsilon} \tag{18}$$

This module is intimately related to the bonding energy of the atoms. A steep slope indicates that large forces are required to separate the atoms and cause the material to deform elastically. Therefore, the material has a high modulus of elasticity. The bonding forces and the modulus of elasticity are generally greater in materials with a high melting point (see **Figure 26**). The module is a measure of the rigidity of the material. A rigid material retains its size and shape [89].

The modulus of resistance (E_r), which is the area that appears under the elastic portion of the stress-strain curve, is the elastic energy that a material absorbs or releases during the application and release of the applied load, respectively.

1. Ductility: it measures the degree of deformation that a material can withstand without breaking. The distance between the calibrated marks in a test tube

before and after the test can be measured. The % elongation represents the distance that the test tube stretches plastically before the fracture:

$$\varepsilon = 100\frac{(l - l_0)}{l_0} \tag{19}$$

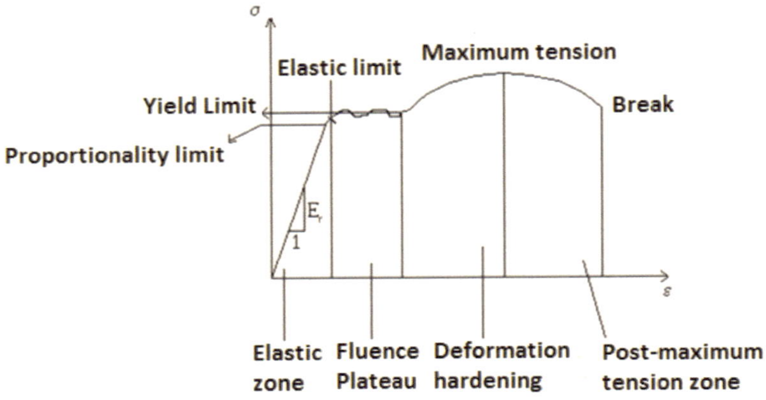

Figure 26.
Deformation diagram.

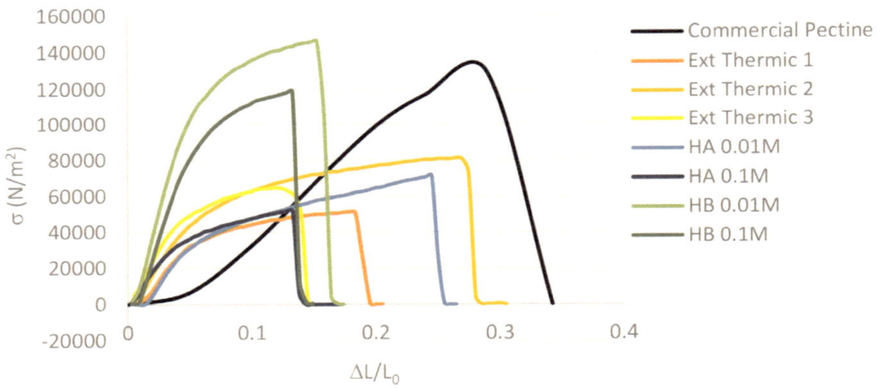

Figure 27.
Mechanical properties of pectin membranes.

Membrane	E (MPa)	ε_{max}	σ_{max} (MPa)
Commercial pectin	6.20	2.90	0.13
HA 0.01 M	0.94	0.24	0.07
HA 0.1 M	0.87	0.14	0.01
HB 0.01 M	2.54	0.15	0.15
HB 0.1 M	2.18	0.13	0.12
Ext. thermic 1	0.97	0.18	0.05
Ext. thermic 2	1.02	0.27	0.08
Ext. thermic 3	1.10	0.11	0.06

Table 4.
Data of mechanical properties of pectin membranes.

where l is the distance between the calibrated marks after the rupture of the material [89].

The commercial pectin membrane has a high value of Young's modulus, but it has good elasticity; this type of very positive characteristics when forming films is mainly due to the most exhaustive and purified purification processes, in which acetyl groups, proteins and nitrogen compounds, polyphenols, and finally sodium ion are eliminated. As for the pectin in which the different hydrolyses were made, the ones with greater rigidity are those made with sodium hydroxide, followed by the thermal extractions and finally the acid hydrolysis, which are the most elastic and flexible [90–94] (see **Figure 27** and **Table 4**).

6. Conclusions

It can be observed through this work that the hydrolyzed pectin obtained from *Citrus sinensis* has inferior properties to commercial pectin. Based on this it can be inferred that the selected procedure requires a greater degree of purification with the addition of pretreatment and posttreatment elimination of polyphenols and other undesired compounds such as proteins, etc. The hydrolyzed pectins that have the best qualities are HA 0.1 M and HA 0.5 M which can be applied to different industries such as pharmaceutical, food, and mining.

Acknowledgements

I would like to especially thank Dr. Ariel Ochoa for providing support in the development of this work (Laboratorio de Investigación y Servicios de Química Física (LISeQF-UNSL), Instituto de Física Aplicada-CONICET-Universidad Nacional de San Luis; Instituto de Tecnología Química, INTEQUI-CONICET-UNSL; Laboratorio de Membranas y Biomateriales INFAP-CONICET-UNSL).

Author details

Paula Ruano[1], Lismet Lazo Delgado[1], Sergio Picco[2], Liliana Villegas[3], Franco Tonelli[4], Mario Eduardo Aguilera Merlo[2], Javier Rigau[5], Darío Diaz[6] and Martin Masuelli[1]*

1 Laboratorio de Investigación y Servicios de Química Física (LISeQF), Instituto de Física Aplicada (INFAP), CONICET-Universidad Nacional de San Luis, San Luis, Argentina

2 Facultad de Ingeniería y Ciencias Agropecuarias (FICA), Departamento de Ingeniería de Procesos, Área de Tecnología de Alimentos, San Luis, Argentina

3 Laboratorio de Microbiología Ambiental, INQUISAL-UNSL-CONICET, San Luis, Argentina

4 Laboratorio de Ingeniería de Reacciones Químicas, FICA-UNSL, San Luis, Argentina

5 Instituto de Tecnología Química (INTEQUI), CONICET-UNSL, San Luis, Argentina

6 Laboratorio de Membranas y Biomateriales INFAP-CONICET-UNSL, San Luis, Argentina

*Address all correspondence to: masuelli@unsl.edu.ar

References

[1] Berk Z. Citrus Fruit Processing. Academic Press is an imprint of Elsevier; 2016

[2] Mohnen D. Pectin structure and biosynthesis. Current Opinion in Plant Biology. 2008;**11**:266-277

[3] Ridley BL, O'Neill MA, Mohnen D. Pectins: Structure, biosynthesis, and oligogalacturonide-related signaling. Phytochemistry. 2001;**57**:929-967

[4] Rolin C. Chapter 10: Pectin. In: Whistler RL, Bemiller JN, editors. Industrial Gums. Polysaccharides and Their Derivatives. 3rd ed. Academic Press; 1993

[5] May CD. Industrial Pectins: Sources, production and applications. Carbohydrate Polymers. 1990;**12**:79-99

[6] Ayora-Talavera T d R, Ramos-Chan CA, Covarrubias-Cárdenas AG, Sánchez-Contreras A, García-Cruz U, Pacheco Lopez NA. Evaluation of pectin extraction conditions and polyphenol profile from *Citrus x lantifolia* waste: Potential application as functional ingredients. Agriculture. 2017;7:28. DOI: 10.3390/agriculture7030028

[7] Cho C-W, Lee D-Y, Kim C-W. Concentration and purification of soluble pectin from mandarin peels using crossflow microfiltration system. Carbohydrate Polymers. 2003;**54**:21-26

[8] Marín FR, Soler-Rivas C, Benavente-García O, Castillo J, Perez-Alvarez JA. By-products from different citrus processes as a source of customized functional fibres. Food Chemistry. 2007; **100**:736-741

[9] Yeoh S, Shi J, Langrish TAG. Comparisons between different techniques for water-based extraction of pectin from orange peels. Desalination. 2008;**218**:229-237

[10] Petersen K, Nielsen PV, Bertelsen G, Lawther M, Olsen MB, Nilssonk NH, et al. Potential of biobased materials for food packaging. Trends in Food Science & Technology. 1999;**10**:52-68

[11] Sanchez-Garcia MD, Lopez-Rubio A, Lagaron JM. Natural micro and nanobiocomposites with enhanced barrier properties and novel functionalities for food biopackaging applications. Trends in Food Science & Technology. 2010;**21**:528-536

[12] Miller KS, Krochta JM. Oxygen and aroma barrier properties of edible films: A review. Trends in Food Science & Technology. 1997;**8**(7): 228-237

[13] Druchta J, De Mulder Johnston C. Edible films solve problems. Food Technology. 1997;**51**:61-74

[14] Anker M. Edible and biodegradable films and coatings for food packaging. Literature Review, SIK-Report no. 623; GoÈteborg, Sweden; 1996

[15] Azeredo HMC, Mattoso LHC, Wood D, Williams TG, Avena-Bustillos RJ, McHugh TH. Nanocomposite edible films from mango puree reinforced with cellulose nanofibers. Journal of Food Science. 2009;**74**(5):N31-N35

[16] Sahin S, Sumnu SG. Physical Properties of Foods. Springer; 2006

[17] Selke SE. Chapter 8: Plastics recycling and biodegradable plastics. In: Handbook of Plastics Technologies. McGraw-Hill; 2006

[18] Chakrabarty T, Kumar M, Shahi VK. Chapter 10: Chitosan based membranes for separation, pervaporation and fuel cell applications: Recent developments. In: Biopolymers. Rijeka, Croatia: IntechOpen; 2010

[19] Mohan T, Kargl R, Doliskaa A, Ehmann HMA, Ribitsch V, Stana-Kleinschek K. Enzymatic digestion of partially and fully regenerated cellulose model films from trimethylsilyl cellulose. Carbohydrate Polymers. 2013; **93**:191-198

[20] Shi A-m, Wang L-j, Li D, Adhikari B. Characterization of starch films containing starch nanoparticles. Part 1: Physical and mechanical properties. Carbohydrate Polymers. 2013;**96**:593-601

[21] Liakos I, Rizzello L, Bayer IS, Pompa PP, Cingolani R, Athanassiou A. Controlled antiseptic release by alginate polymer films and beads. Carbohydrate Polymers. 2013;**92**:176-183

[22] Li R, Hu P, Ren X, Worley SD, Huang TS. Antimicrobial N-halamine modified chitosan films. Carbohydrate Polymers. 2013;**92**:534-539

[23] Saurabh CK, Gupta S, Bahadur J, Mazumder S, Variyar PS, Sharma A. Radiation dose dependent change in physiochemical, mechanical and barrier properties of guar gum based films. Carbohydrate Polymers. 2013;**98**: 1610-1617

[24] Heikkinen SL, Mikkonen KS, Pirkkalainen K, Serimaa R, Joly C, Tenkanen M. Specific enzymatic tailoring of wheat arabinoxylan reveals the role of substitution on xylan film properties. Carbohydrate Polymers. 2013;**92**:733-740

[25] Tajik S, Maghsoudlou Y, Khodaiyan F, Jafari SM, Ghasemlou M, Salami M. Soluble soybean polysaccharide: A new carbohydrate to make a biodegradable film for sustainable green packaging. Carbohydrate Polymers. 2013;**97**: 817-824

[26] Jouki M, Khazaei N, Ghasemlou M, Nezhad MH. Effect of glycerol concentration on edible film production from cress seed carbohydrate gum. Carbohydrate Polymers. 2013;**96**:39-46

[27] Espitia PJP, Du WX, de Jesús Avena-Bustillos R, Soares NDFF, McHugh TH. Edible films from pectin: Physical-mechanical and antimicrobial properties—A review. Food Hydrocolloids. 2014;**35**:287-296

[28] Parris N, Coffin DR, Joubran RF, Pessen H. Composition factors affecting the water vapor permeability and tensile properties of hydrophilic films. Journal of Agricultural and Food Chemistry. 1995;**43**:1432-1435

[29] Pasini Cabello SD, Takara EA, Marchese J, Ochoa NA. Influence of plasticizers in pectin films: Microstructural changes. Materials Chemistry and Physics. 2015;**162**: 491-497

[30] Azeredo HMC, Morrugares-Carmona R, Wellner N, Cross K, Bajka B, Waldron KW. Development of pectin films with pomegranate juice and citric acid. Food Chemistry. 2016;**198**: 101-106

[31] Fishman ML, Chau HK, Qi PX, Hotchkiss AT, Garcia RA, Cooke PH. Characterization of the global structure of low methoxyl pectin in solution. Food Hydrocolloids. 2015;**46**:153-159

[32] Abid M, Cheikhrouhou S, Renard CMGC, Bureau S, Cuvelier G, Attia H, et al. Characterization of pectins extracted from pomegranate peel and their gelling properties. Food Chemistry. 2017;**215**:318-325

[33] Kurita O, Fujiwara T, Yamazaki E. Characterization of the pectin extracted from citrus peel in the presence of citric acid. Carbohydrate Polymers. 2008;**74**: 725-730

[34] Venzon SS, Canteri MHG, Granato D, Junior BD, Maciel GM,

Stafussa AP, et al. Physicochemical properties of modified citrus pectins extracted from orange pomace. Journal of Food Science and Technology. DOI: 10.1007/s13197-014-1419-2

[35] Ying R, Rondeau-Mouro C, Barron C, Mabille F, Perronnet A, Saulnier L. Hydration and mechanical properties of arabinoxylans and β-d-glucans films. Carbohydrate Polymers. 2013;**96**:31-38

[36] Cerqueira MA, Souza BWS, Teixeira JA, Vicente AA. Effect of glycerol and corn oil on physicochemical properties of polysaccharide films. A comparative study. Food Hydrocolloids. 2012;**27**:175-184

[37] Narasimha Murthy S, Hiremath SRR, Paranjothy KLK. Evaluation of carboxymethyl guar films for the formulation of transdermal therapeutic systems. International Journal of Pharmaceutics. 2004;**272**: 11-18

[38] Lasko CL, Hurst MP. An investigation into the use of chitosan for the removal of soluble silver from industrial wastewater. Environmental Science & Technology. 1999;**33**: 3622-3626

[39] Thakhiew W, Devahastin S, Soponronnarit S. Effects of drying methods and plasticizer concentration on some physical and mechanical properties of edible chitosan films. Journal of Food Engineering. 2010;**99**: 216-224

[40] Yang L, Paulson AT. Effects of lipids on mechanical and moisture barrier properties of edible gellan film. Food Research International. 2000;**33**: 571-578

[41] Rao MS, Kanatt SR, Chawla SP, Sharma A. Chitosan and guar gum composite films: Preparation, physical, mechanical and antimicrobial properties. Carbohydrate Polymers. 2010;**82**:1243-1247

[42] Maya JJ, Sabu T. Biofibres and biocomposites. Carbohydrate Polymers. 2008;**71**:343-364

[43] Pilgrim GW. Jams, jellies and preserves. In: The Chemistry and Technology of Pectin, Food Science and Technology Series of Monographs. 1991. pp. 23-50

[44] Ordoñez FJM. Extracción y Caracterización de la Pectina Obtenida a Partir del Fruto de dos Ecotipos de Cocona (*Solanum sessiliflorum*), en Diferentes Grados de Madurez; a Nivel de Planta Piloto. Universidad Nacional de Colombia; 2011

[45] McCabe WL, Smith JC, Harriott P. Operaciones Unitarias en Ingeniería Química. 4th ed. México; 2005

[46] Rodriguez F, Aguado J, Calles JA, Cañizares P, López B, Santos A, et al. Ingeniería de la Industria Alimentaria, Volumen II. Operaciones de Procesado de Alimentos. España: Editorial Síntesis; 2002

[47] Whistler RL, BeMiller JN. Alkaline degradation of polysaccharides. In: Advances in Carbohydrate Chemistry. Vol. 13. Academic Press; 1958. pp. 289-329

[48] Joye DD, Luzio GA. Process for selective extraction of pectins from plant material by differential pH. Carbohydrate Polymers. 2000;**43**(4): 337-342

[49] Granados Cristancho YA. Importancia de los Ensayos TGA y DSC en el Estudio de las Propiedades Térmicas de Mezclas Asfálticas. Bogotá, Colombia; 2017

[50] Brown ME, Gallagher PK. Handbook of Thermal Analysis and Calorimetry: Recent Advances,

Techniques and Applications. Vol. 5. Elsevier; 2011

[51] Wagner M. Thermal Analysis in Practice. Mettler Toledo; 2010

[52] Wang X, Chen Q, Lü X. Pectin extracted from apple pomace and citrus peel by subcritical water. Food Hydrocolloids. 2014;**38**:129-137

[53] Kowalski G, Kijowska K, Witczak M, Kuterasinski Ł, Łukasiewicz M. Synthesis and effect of structure on swelling properties of hydrogels based on high methylated pectin and acrylic polymers. Polymers. 2019;**11**:114

[54] Sutar PB, Mishra RK, Pal K, Banthia AK. Development of pH sensitive polyacrylamide grafted pectin hydrogel for controlled drug delivery system. Journal of Materials Science: Materials in Medicine. 2008;**19**: 2247-2253

[55] Miceli-Garcia LG. Pectin from apple pomace: Extraction, characterization, and utilization in encapsulating alpha-tocopherol acetate [Master Science thesis]. Lincoln, Nebraska, USA: University of Nebraska – Lincoln; 2014

[56] Lim J, Yoo J, Ko S, Lee S. Extraction and characterization of pectin from Yuza (*Citrus junos*) pomace: A comparison of conventional-chemical and combined physical-enzymatic extractions. Food Hydrocolloids. 2012; **29**:160-165

[57] Ravindrakullai Reddy M, Manjunath K. Evaluation of pectin derived from orange peel as a pharmaceutical excipient. International Journal of Drug Development and Research. 2013;**5**(2):283-294

[58] Ramana G, Krishna Chaitanya A. Preparation and In-vitro characterization of ethylcellulose coated pectin alginate microspheres of 5-fluorouracil for colon targeting. Journal of Applied Pharmaceutical Science. 2011;**1**(8):170-176

[59] Calce E, Bugatti V, VV, De Luca S. Solvent-free synthesis of modified pectin compounds promoted by microwave irradiation. Molecules. 2012; **17**:12234-12242

[60] Krapivnitska IO, Gursky PV, Pertsevy FV. Thermogravimetric analysis of pectin gel. ScienceRise. 2015; 7(2):23-28

[61] Kazemzadeh B, Hosseinzadeh H, Babazadeh M. Synthesis of a novel pectin-based superabsorbent hydrogel with salt and pH-responsiveness properties. Biomedical and Pharmacology Journal. 2013;**6**(1):41-48

[62] Zawieja Z, Sawicki J. Thermogravimetric analysis of modified cellulose mixtures. Archives of Metallurgy and Materials. 2019;**64**(1): 153-157

[63] Kalathaki I, Alba, K, Muhamedsalih H, and Kontogiorgos V. Fabrication and characterisation of metal-doped pectin films. Food Hydrocolloids. In Press; 2019

[64] Shi L, Gunasekaran S. Preparation of pectin–ZnO nanocomposite. Nanoscale Research Letters. 2008;**3**: 491-495

[65] Zaleska H, Ring SG, Tomasik P. Apple pectin complexes with whey protein isolate. Food Hydrocolloids. 2000;**14**:377-382

[66] Aburto J, Moran M, Galano A, Torres-García E. Non-isothermal pyrolysis of pectin: A thermochemical and kinetic approach. Journal of Analytical and Applied Pyrolysis. 2015; **112**:94-104

[67] Wang W, Ma X, Jiang P, Hu L, Zhi Z, Chen J, et al. Characterization of pectin from grapefruit peel: A

comparison of ultrasound-assisted and conventional heating extractions. Food Hydrocolloids. 2016;**61**:730-739

[68] Xu Y, Zhang L, Bailina Y, Ge Z, Ding T, Ye X, et al. Effects of ultrasound and/or heating on the extraction of pectin from grapefruit peel. Journal of Food Engineering. 2014;**126**:72-81

[69] Einhorn-Stolla U, Kunzeka H, Dongowski G. Thermal analysis of chemically and mechanically modified pectins. Food Hydrocolloids. 2007;**21**: 1101-1112

[70] Methacanon P, Krongsin J, Gamonpilas C. Pomelo (*Citrus maxima*) pectin: Effects of extraction parameters and its properties. Food Hydrocolloids. 2014;**35**:383-391

[71] Gómez R, Murillo R. Espectroscopia Infrarroja. Universidad Nacional Autónoma de México; 2006

[72] Acikgoz C. Extraction and characterization of pectin obtained from quince fruits (*Cydonia vulgaris* pers) grown in Turkey. Asian Journal of Chemistry. 2011;**23**(1):149-152

[73] Kamnev AA, Colina M, Rodriguez J, Ptitchkina NM, Ignatov VV. Comparative spectroscopic characterization of different pectins and their sources. Food Hydrocolloids. 1998; **12**(3):263-271

[74] Lima MS, Paiva EP, Andrade SAC, Paixão JA. Fruit pectins—A suitable tool for screening gelling properties using infrared spectroscopy. Food Hydrocolloids. 2010;**24**(1):1-7

[75] Ferreira D, Barros A, Coimbra MA, Delgadillo I. Use of FT-IR spectroscopy to follow the effect of processing in cell wall polysaccharide extracts of a sun-dried pear. Carbohydrate Polymers. 2001;**45**(2):175-182

[76] Kalapathy U, Proctor A. Effect of acid extraction and alcohol precipitation conditions on the yield and purity of soy hull pectin. Food Chemistry. 2001; **73**(4):393-396

[77] Baum A, Dominiak M, Vidal-Melgosa S, Willats WGT, Søndergaard KM, Hansen PW, et al. Prediction of pectin yield and quality by FTIR and carbohydrate microarray analysis. Food and Bioprocess Technology. 2017;**10**:143-154

[78] Kyomugasho C, Christiaens S, Shpigelman A, Van Loey AM, Hendrickx ME. FT-IR spectroscopy, a reliable method for routine analysis of the degree of methylesterification of pectin in different fruit- and vegetable-based matrices. Food Chemistry. 2015; **176**:82-90

[79] Sato M d F, Rigoni DC, Canteri MHG, Petkowicz CL d O, Nogueira A, Wosiacki G. Chemical and instrumental characterization of pectin from dried pomace of eleven apple cultivars. Acta Scientiarum. Agronomy. 2011;**33**(3):383-389

[80] Fernández IJ. Polímeros en solución y aplicación de los polímeros en la industria petrolera. Revista Iberoamericana de Polímeros. 2003; **4**(2):1-14

[81] Masuelli MA. Viscometric study of pectin. Effect of temperature on the hydrodynamic properties. International Journal of Biological Macromolecules. 2011;**48**:286-291

[82] Masuelli MA. Mark-Houwink parameters for aqueous-soluble polymers and biopolymers at various temperatures. Journal of Polymer and Biopolymer Physics Chemistry. 2014; **2**(2):37-43

[83] Rubio-Anaya M, Guerrero-Beltrán JA. Polímeros utilizados para la elaboración de películas biodegradables.

Temas Selectos de Ingeniería de Alimentos. 2012;**6**:173-181

[84] Walter RH, Taylor S, editors. The Chemistry and Technology of Pectin. Food Science and Technology. USA: Academic Press; 2012

[85] Thakur BR, Singh RK, Handa AK, Rao DMA. Chemistry and uses of pectin—A review. Critical Reviews in Food Science and Nutrition. 1997;**37**(1): 47-73

[86] Tyagi V, Sharma PK, Malviya R. Pectins and their role in food and pharmaceutical industry: A review. Journal of Chronotherapy and Drug Delivery. 2015;**6**(3):65-78

[87] Ipohorski M, Bozzano PB. Microscopía Electrónica de Barrido en la Caracterización de Materiales. Argentina: CNEA; 2002

[88] Siracusa V. Food packaging permeability behaviour: A report. International Journal of Polymer Science. 2012;**2012**

[89] Askeland DR. Cap. 7: Endurecimiento por deformación y recocido. In: Ciencia e Ingeniería de los Materiales. 3rd ed. International Thomson; 1998

[90] Masuelli MA. Chapter 1: Introduction of fibre-reinforced polymers—Polymers and composites: Concepts, properties and processes. In: Fiber Reinforced Polymers—The Technology Applied for Concrete Repair. Rijeka, Croatia; 2013. DOI: 10.5772/54629

[91] Masuelli MA, García MG. Biopackaging: Tara Gum Films. In: Masuelli M, Renard D, editors. Advances in Physicochemical Properties of Biopolymers. Part 2. Abu Dhabi, UAE: Editorial Bentham Publishing; 2017. pp. 876-898

[92] Martín A, Masuelli MGG, Marchese J, Ochoa NA, et al. Journal of Applied Polymer Science. 2015;**132**: 41624. DOI: 10.1002/app.41624. ISSN: 1097-4628

[93] Masuelli MA. Alcayota films: Effect of crosslinking. Journal of Advanced Chemical Engineering. 2016;**6**(3):49. DOI: 10.4172/2090-4568.C1.002

[94] Zanon M, Masuelli M. Alcayota gum films: Experimental reviews. Journal of Materials Science and Chemical Engineering. 2018;**6**:11-58. DOI: 10.4236/msce.2018.65002

Chapter 2

Extraction and Purification of Pectin from Agro-Industrial Wastes

Erumalla Venkatanagaraju, N. Bharathi, Rachiraju Hema Sindhuja, Rajshree Roy Chowdhury and Yarram Sreelekha

Abstract

With the advent of science and technology, agro-industrial wastes are converted into various value-added products to meet the demands of increasing population. In recent years, natural polymers have evoked tremendous interest due to easy conversion into value-added products. Apart from various natural polymers, pectin occupied a prominent place due to diverse pharmaceutical and therapeutic applications. Excess utilisation of pectin, the gap between production and demand is widening. To fulfil this gap various techniques are adopted for obtaining high yield pectin from various agro-industrial wastes. This chapter will be focusing on extraction and purification of pectin from various agro-industrial wastes, considered as main environmental pollutants.

Keywords: agro-industrial, waste, pectin, pharmaceutical, therapeutic

1. Introduction

Pectins are complex branched polysaccharides present in the primary cell wall of plants [1]. It is a highly valued food ingredient commonly used as a gelling agent and stabilizer [2]. It is usually extracted by chemical or enzymatic methods from fruits [3]. Pectin is considered as the most complex macromolecule in nature, since it can be composed of up to 17 different monosaccharides containing more than 20 different linkages [4].

Pectins are enriched with repeated units of methyl ester galacturonic acid [4]. They form chemically stable and physically strong skeletal tissues of plants when combined with proteins and other polysaccharides [5]. They are usually produced in the initial stages of primary cell wall growth and make one third of the cell wall in both monocots and dicots [6]. Pectin is significantly reduced or absent in non-extendable secondary cell walls and is the only major class of plant polysaccharide largely limited to primary cell walls [7]. Pectin imparts strength and flexibility to the cell wall, apart from number of fundamental biological functions such as signalling, cell proliferation, differentiation, cell adhesion and maintaining turgor pressure of cell [8]. Pectins are involved in regulating mobility of water and plant fluids through the rapidly growing parts [6]. It also influences the texture of fruit and vegetables [9]. Apple pomace and orange peel are the two major sources of commercial pectin due to the poor gelling behaviour of pectin from other sources [6].

Pectin is one of the most important polysaccharides due to its increasing demand in the global market, reaching a total production capacity of around 45–50 Million tonnes per annum. While the demand in 2011 was approximately 140–160 Million tonnes per annum, earning the interest of industry in this complex polysaccharide processing [10]. Pectins have received considerable attention as a high fibre diet that benefits health by reducing cholesterol and, serum glucose levels and acting as anticancer agents [11]. Pectins have shown promising results as drug carriers for oral drug delivery and are widely used for various bio-medical applications [5]. In addition, pectin has been described as an emerging prebiotic with the ability to modulate colon microbiota [12]. Considering above properties and applications, pectin has gained immense priority in the global biopolymer market with great potential and opportunities for future developments.

2. Structure and properties of pectin

One of the most abundant macromolecules present in the primary cell wall of the plants is pectin; their presence is detected in the matrix as well as in the middle lamellae [7]. Pectin is highly rich with galacturonic acid (GalA), that forms the backbone of three more domains found along with pectin that are homogalacturonan (HGA), rhamnogalacturonan-I (RG-I) and rhamnogalacturonan-II (RG-II) [13]. About 70% of pectin is mainly composed of galacturonic acid (GA) [14]. Pectin is made of three polysaccharides that are covalently linked together, thus forming pectin networks in the cell wall matrix and the middle lamellae [15, 16].

Homogalacturonan (HG) takes up about 60–65% of the total pectin [3, 17], with a backbone of alpha-1,4-linked GalA residues, these GalA residues are methyl esterified which has an important role in the physical properties of pectin [4]. The presence of HG is seen to be present in approximately 100 GalA residues, but there are cases when its detected interspersed within other pectin polysaccharide [14]. On the other hand rhamnogalacturonan-I (RG-I) backbone which contributes 20–35% of pectin is composed of repeated and alternating groups of L-rhamnosyl and D-galacturonosyl residues [18]. There can be as many repeats as 300 of this disaccharide in case of sycamore cells, which are cultured in suspension [3, 16, 19]. The rhamnosyl residues have side chains of sugars which are mainly consisting of either galactosyl or arabinosyl residues [5]. The GalA residue of RGI unlike HGA are mostly not methyl esterified [20].

Rhamnogalacturonan-II (RG-II) is one of the highly conserved and complex structure which consist of distinct regions within HG, which makes up about 10% of the pectin [3], they have side chains of four different types with a particular sugar residue like aceric acid, apiose-3-deoxy-lyxo-2-heptulosaric acid, and 3-deoxy-manno-2-octulosonic acid. The HG residues along with nine of the GalA residues are attached to these side chains [3, 5]. There are other substituted HG residues that make up pectin such as xylogalacturonan and apiogalacturonan whose expression is restriction. Even a minor mutation in R-II structure can lead to defects in the plant growth like dwarfism, thus suggesting its importance for normal growth of plant [3]. RG-I being highly branched in nature thus, called as the hairy region of pectin on the other hand HGA domain are known as the smooth region [7]. It is generally believed and noticed that there is covalent linkage within the pectin polysaccharides and pectin degrading enzymes are needed to separate and isolate HG, RG-I and RG-II from each other [21, 22]. Due to their similarity in

HG and RG-II backbone structure composing of 1-4-linked alpha-D-GalA residues, they are likely to be linked covalently but there are no reports of RG-I to be covalently linked with HG [23].

3. Properties of pectin

3.1 Colloidal

Pectin precipitates as a solid gel on treating with a dehydrating agent like alcohol. They are extremely sensitive to dehydration and get effected by any other hydrophilic colloids as well, thus they are known to be insoluble in most of the bio-colloids. The negative charge of pectin depends on the number of free carboxyl group that is mainly responsible for its precipitation [24].

3.2 Solubility

Based on solubility pectins are of two types i.e., water soluble and water insoluble. Factors affecting the solubility of pectin are pH, temperature, nature of the solute and concentration of the solute [6, 13]. Pectin attains stability at a pH of 4 [17]. The solubility of pectin also depends on its composition like monovalent cation of pectin are soluble in water whereas di or trivalent are insoluble in water.

3.3 Gelation

One of the most interesting properties of pectin is its ability to form gel in the presence of either acid or calcium or sugar, this enables them to be used in many food industries [15]. Hydrogen bonding and hydrophobic interactions between polymer chains stabilizes the pectin polymer [9].

4. Extraction and purification of pectin from agrowaste

Pectin is a high molecular weight polysaccharide that is present in almost all plants and help in maintaining the integrity of cell structure. Pectin is used in food industries to increase the viscosity of food products such as beverages, jams and jellies. It also has implications in pharmaceutical industry, especially in drug formulations, as an excipient due to its characteristics in release kinetics. Due to increased demand for pectin in food, pharmaceutical and therapeutic applications, thus, require efficient extraction processes. In order to increase the yield of pectin, various extraction methods have been adapted to obtain insoluble pectin present in the middle lamellae of plant cells, one of them being heating in acidic medium that makes insoluble pectin as soluble. Ripening of fruits also converts insoluble pectin into soluble pectin. Pectin can be extracted from various kinds of fruits, but the most commercial form of pectin is extracted from the peels of citrus fruits by alcohol precipitation [9, 25]. Citrus fruits contain 0.5–3.5% pectin which is largely present in the peel of fruits [26].

Pectin has been isolated from various plants such as apple [27], citrus peel, carrots [28], sugar beet pulp [29, 30], sunflower heads [31], papaya [32] and oranges [33]. The most commonly used method for extracting pectin from plant tissue is by heating the plant sample in acidified water. The addition of extra chelating agents such as EDTA to the extraction mixture helps in easy release of pectin from cell wall. Care should be taken not to perform a long period of direct heating as it may lead to the thermal

Material	Extraction process	Pectin (%)	References
Cacao pod husk (*Theobroma cacao*)	Acid extraction	3.7–8.6	[34]
Mangosteen rind (*Garcinia mangostana*)	Chemical treatment	23.5	[35]
Durian rind (*Durio zibethinus*)	Acid extraction	2.1–10.25	[36]
Orange peels (*Citrus sinensis*)	Acid extraction	0.2–6	[37]
Lemon peels (*Citrus limon*)	Acid extraction	0.8–8	[37]
Dragon fruit peels (*Hylocereus undatus*)	Ultrasound assisted	1.89–7.65	[38]
Banana-stem, leaf, peel (*Musa acuminata*)	Alcohol precipitation	4–13.8	[39]
Orange peel	Alcohol precipitation	7.9	[39]
Cucumis melo	Aqueous acid extraction alcohol precipitation	4.53	[40]
Cocoa peel	Microwave assisted	42.3	[41]
Apple Pomace	Acid extraction	12.9–20.9	[27]
Lime-peel and pulp	Microwave assisted extraction under pressure	8–17.9	[42]
Watermelon rind	Acid and enzymatic extraction		[43]
Orange peels	Acid extraction	5.4–26.3	[44]
Sweet potato peels	Acid extraction	2.59–5.08	[45]
Orange peel	Ultrasound assisted	20.92	[46]
Orange peels (*Citrus sinensis*)	Acid extraction	29.41	[47]
Kaffir lime peel (*Citrus hystrix*)	Chemical and acid extraction	61.8	[48]
Punica granatum peels	Acid extraction	27	[49]
Orange peel (*Citrus sinensis*)	Acid extraction	45.5	[50]
Lemon (*Citrus limon*)	Acid extraction	2.7–16.7	[51]
Orange (*Citrus sinensis*)	Acid extraction	1.6–15.9	[51]
Grape (*Citrus paradisi*)	Acid extraction	2.3–15.7	[51]
Orange peel	Water-based extraction	2.2	[52]
Sweet potato peel (*Ipomoea batatas*)	Alkaline extraction	16.78	[45]
Tomato waste	Ultrasound assisted	15.1–35.7	[53]
Pumpkin peels	Soxhlet extraction	6.8–7.7	[54]
Lemon pomace	Acid extraction	10.3–13.1	[55]
Jackfruit waste (*Artocarpus heterophyllus Lam*)	Acid and chemical extraction	12–15	[27]
Lemon peel wastes	Aqueous extraction		[56]
Citric waste	Acid extraction	78	[57]
Apple peel waste (*Malus pumila. Cv Amri*)	Acid and chemical extraction	1.21	[58]
Horse eye bean peel (*Mucuna urens*)	Acid extraction	4.4	[59]
Banana peel	Acid extraction	11.31	[60]
Mango peel	Acid extraction	18.5	[60]
Grapefruit peel	Alcohol extraction	25	[33]

Material	Extraction process	Pectin (%)	References
Saba banana peel (*Musa acuminata* × *Musa balbisiana*)	Acid extraction	17.05	[61]
Passion fruit peels	Acid extraction	2.25–14.6	[62]
Citrus peel	Acid extraction	25.5	[63]
Pumpkin waste (*Cucurbita maxima*)	Acid hydrolysis	2.90	[39]
Mango peel	Acid extraction	20.8	[64]
Jackfruit wastes (*Artocarpus integer*)	Optimised acid extraction	38.42	[65]
Citrus depressa endocarp	Acid extraction	4.1	[66]
Orange peel	Acid extraction	7.3–52.9	[67]
Jackfruit waste (*Artocarpus heterophyllus*)	Chemical and acid extraction	8.9–15.1	[27]

Table 1.
Methods of extraction of pectin from various agrowaste compounds.

degradation of the polymer. Extraction process of pectin is carried out under reflux using acidified water at 97°C for 30 min. The hot acid extract was then filtered using a cheese cloth to remove the pulp. The filtrate was then cooled to 4°C and precipitated using double the volume of ethanol. The solvent precipitate mixture is then mixed till the pectin floats and removed by using cheese cloth followed by drying [27].

Pectin is also extracted from dried sugar beet pulp after treating with acidified medium followed by purification through alcohol precipitation. Xin Huang et al., slightly modified the traditional method, where the sample was diluted with deionized water and was made acidic (pH −1.2) by using HCl. The sample was then heated to 90°C for 3 h and cooled to 40°C (pH −4.5) with 25 g/100 g ammonia. The mixture was then filtered using a Buchner funnel and pectin was precipitated using ethanol [29]. The ethanol is removed by squeezing with nylon cloth and washed several times followed by drying.

The carrot pomace is also used for pectin extraction by treating with hot aqueous citric acid solution adjusted to the desired pH. The pectin yield was maximum at the following optimum conditions: pH −1.3; temperature 90°C; heating time 79.8 min. Under these conditions, the extraction yield of carrot pomace pectin was found to be 16.0%. The extract mixture was then allowed to cool, filtered and precipitated by using ethanol in the ratio 2:1 [28]. Dried papaya peel can be used in pectin extraction where the majority of the lipids, and soluble pigments are removed by treating with ethanol and acetone. It is repeatedly homogenised with 95% ethanol and filtered until the filtrate becomes clear. The final filtration was done with the residue homogenised in acetone and dried overnight to obtain the alcohol insoluble residue (AIR). The majority of the pectin in the papaya AIR is present as chelator soluble pectin (CSP) followed by sodium carbonate soluble pectin (SSP) and water-soluble pectin (WSP). The WSP fraction is first obtained from the AIR by boiling it in water and filtering it. The remaining residue is treated with 0.05 M cyclohexane trans-1,2-diamine tetra-acetic acid (CDTA) in 0.1 M potassium acetate (pH 6.5) for 6 h at 28°C and filtered to give the CSP fraction. The residue is then treated with 0.05 M sodium carbonate solution having 0.02 M $NaBH_4$ for 16 h at 4°C, and subsequently for 6 h at 28°C. The solution when filtered gives the SSP fraction of the AIR [32].

Agro-industrial wastes can be used as the raw material for the production of industrial low and high methoxy pectin. The alcohol insoluble material (AIM) produced from dried agrowaste by boiling it with 3 volumes of ethanol for 25 min and continuous washing with 70% ethanol to remove impurities such as pigments, free sugars, etc. Sunflower heads also act as potential sources for pectin extraction. The heads are washed by hot distilled water through a mesh or cheese cloth and the pectin was precipitated by addition of 1 M nitric acid at 1:5 acid:filtrate ratio [34]. The mixture was maintained for 1 h at 5°C and was washed six times in ethanol solvent at 1:2 gel:solvent ratio to remove the impurities and to increase pH by removing the acid [31]. The washed pectin gel can be dried in a vacuum oven at 55°C for 16 h. The dried pectin flakes are ground into a powder for further use (**Table 1**).

5. Characterization

Large amounts of fruit wastes that are being generated can be disposed effectively by manufacturing beneficial by-products like pectin. Pectin is used to increase foaming power of gases, as agglutinator, as filler in pharmaceutical preparations and also in food industry. The use of pectin for different purposes depends on its characters like acetyl value, degree of esterification, uronic acid content and methoxyl content, etc. [68]. The amount of anhydrouronic acid gives the purity of pectin which is not less than 65% for pectin that is used commercially [69].

5.1 Qualitative tests

Colour reader can be used to measure the colour of extracted pectin by placing lens of reader on sample powder. The colour of the extracted samples can be compared with that of commercial pectin. Solubility of pectin in different solvents is measured i.e., solubility in cold and, hot water and alkali like NaOH.

5.2 Quantitative tests

Acetyl content and equivalent weight of pectin can be estimated using NaOH whereas methoxyl content can be estimated by saponification and titration. Degree of esterification can be calculated from methoxyl content and anhydrouronic acid content. After acid hydrolysis, sugar separation can be achieved by thin layer chromatography. Intrinsic viscosity of pectin is measured by dissolving pectin in water and by preparation of solutions of various concentrations [27, 32, 60].

6. Applications

Pectin being a great inert, biodegradable and biocompatible complex, is widely used in various fields such as in textiles, food industries as gelling agents, pharmaceuticals and other products as well [70]. Pectin are used as biomaterials in gene delivery [71], application in oral drug delivery [72], as edible coating for food packaging [25], biomass yield and bio refinery [21, 22]. It also has applications in tissue engineering as scaffolds [73], in paper and textile industries for the preparation of ultracentrifugation membrane [74].

6.1 As food product

From the very early period of time, pectin had become one of the major natural constituents of human food, and they have been widely used as a gelling agent for jams and jellies. In jam processing, fruits are cooked properly in order to release juice and pectin which converts the proto-pectin into soluble pectin [19]. Pectins are also used as a substitute of sugar in jams that are made without sugar, using LM (low methoxy) pectin due to its stability in acidic condition. Pectin is widely used for making instant jellies for bakery production these are made with the use of HM (high methoxy) pectin that are thermally stable, the only difference between HM and LM pectin is the amount of pectin in the formula, LM requires a higher amount than that of HM [75]. Other food products like artificial cherries [15], are used to make different kinds of gel puddings that is made of pectin present in the fruit syrup and cold milk [25]. Edible coating of food material is also made of pectin [25]. Pectin is used in beverages as a beverage clouding agent like in diabetic soft drinks [76]. Pectins are also used in the fruit preparation of yogurt to make it more soft and to obtain the partial gel texture [77].

6.2 Biomedical application

6.2.1 Pectin as films

The blending of the natural and synthetic polymers is one of the promising areas of development, this gives new polymeric material with better durability and resistance. Materials like sponge, hydrogels, encapsulating drugs etc. are produced by polymer films [32]. Due to development and discovery of natural polymers scientist have started to form bio-based material rather than synthetic one due to its physiochemical properties like biodegradability, this shift is mainly caused due to the environmental issues and concern regarding the heavy use of plastic [78]. Films of pectin are used to encapsulate and thicken food, and in pharmaceuticals [29]. Hoagland et al. made pectin films with glycerol and lactic acid to prevent fungal contamination on the laminated films [47]. The similar kind of products were made by Fisherman et al., where an edible pectin blend film were plasticized with glycerol, they also suggested that the glass transition at about 50°C was large in case of pectin films, which indicates that the films were fairly flexible at room temperature [79]. Liu et al., made different varieties of biofilms one each with pectin, fish skin gelatin and soybean flour protein which in turn resulted in a composite film that showed an increase in stiffness as well as the strength, whereas decrease in water solubility and water vapour transmission rate when compared to the film that was made with pectin alone. They thus suggested that the tensile strength can be improved by crosslinking the films with methanol or glutaraldehyde [80]. A bio-reactive substance for tissue regeneration was developed by Liu et al., which was composed of pectin or PGLA matrix, which demonstrated that pectin was able to carry signals to molecules, further they suggested that the pectin also helped in the adhesion of the cell and promotes cell proliferation [35]. Some researches have reported the use of pectin membranes as a wound dressing material [81].

6.2.2 Drug delivery

In recent years, biomedical application especially in case of drug delivery system, the use of natural polymers is preferred over the other types due to their inert nature and its biocompatibility. Pectin as the natural polymer is a new developed interest for drug delivery application due to its properties of gel formation in acidic condition, its mucoadhesiveness and its ability to dissolve in basic environment [36]. These properties of pectin are applied in different ways such as the mucoadhesiveness helps in targeting and controlling the drug delivery especially in the nasal and gastric environment, where as its ability to dissolve in basic condition helps in the release of colon related drugs and the formation of gel helps in increasing the contact time of drug in gastric condition [35, 36]. Recent studies have shown the use of LM pectin for nasal drug delivery due to its mucoadhesive property they have a tendency to bind to the mucin with the help of hydrogen bond [82]. Its use in the production of fentanyl (painkiller) has also been seen that help in treating cancer pain which needs rapid drug release [83, 84]. An alternative for smoking cessation are the nasal pectin containing nicotine [85]. As pectin have resistance towards proteases and amylases it has been highly preferred as an encapsulating nanoparticle for drug delivery as most of the proteins are easily degraded by our digestive enzymes and thus to protect these drugs the use of pectin as an outer cover that cannot be degraded in the gastrointestinal tract are preferred for colon and oral drugs [86]. Studies have shown that pectin is able to inhibit cancer metastasis and primary tumour growth in many animal related cancer [87, 88]. Gal-3 is one of the important factors controlling cancer progression and metastasis, and pectin has the ability to recognise these Gal3 components [89]. In a study, citrus pectin was used to target Gal3 that could inhibit the metastatic successfully [87, 90].

6.2.3 As gene delivery and nanoparticles

The treatment of any genetic disorder is called gene therapy as it deals with the defected genes that are responsible for the disorder; these are treated by replacing the defective gene, silencing the unwanted gene expression or by substituting missing genes and these are carried out with the help of viral or non-viral vectors [36]. The use of non-viral vectors is preferred over viral due many reasons like biocompatibility, minimal toxicity and immunogenic reactions of our body [71]. These non-viral vectors are made of polymers of polycationic, chitosan or even pectin. It has been observed that the use of carbohydrate mediated products have better binding capacity, to facilitate the uptake by target cell [91–93]. Pectins were found to be suitable as a coating substance for b-PEI [94, 95]. Opanasopit et al. has also observed the formation of pectin nanoparticle which in turn helps to entrap the DNA for transfection [96]. Katav et al., modified pectin with the help of three different amine groups and these complexes were able to bind with plasmid DNA and there efficiency to transfect or their potential as a non-viral gene delivery carrier was compared and suggested that modified pectin has a promising role in gene delivery [71]. Similar type of study was conducted by Opanaopit et al., where pectin ability as a nanoparticle for gene delivery were studied and the study suggested the potential use of pectin as delivery vector to be safe [96]. Pectin has also been used as wound dressing material in the form of pectin-chitosan based nanoparticles. It has the ability to create an acidic environment in which the bacteria cannot grow. Burapapadh et al. developed a pectin based nanoparticle to improve and enhance the drug dissolution of ITZ (Itraconazole) [97].

6.2.4 Pectin-based scaffolds

Scaffolds are 3-D biomaterials that are porous in nature and are designed to be applied in various fields, few of its basic functions are to promote cell adhesion, to allow enough nutrients and gases transportation and mainly for tissue engineering [32]. Tissue engineering mainly involves the use of biocompatible scaffolds materials to act as a support matrix or to be used as a substrate for delivery of some compounds. There has been a great research going on to promote tissue reconstructions. Coimbra et al., prepared pectin based scaffolds to be used for bone tissue engineering [98]. Similar study was performed by Munarin et al., who examined the use of pectin as injectable biomaterial for bone tissue engineering [99]. Ninan et al. were also able to fabricate biopolymer scaffold of pectin and other compounds using the technique of lyophilisation, thus suggested the use of pectin as ideal polymeric matrix for tissue engineering [73, 100, 101].

DOI: http://dx.doi.org/10.5772/intechopen.85585

7. Conclusion

Pectin is one of the most extensively studied natural biodegradable polymer. In spite of its availability in a large number of plant species, commercial sources of pectin are very limited. There is, therefore, a need to explore other sources of pectin or modify the existing sources to obtain pectin of desired quality attributes. Current knowledge of the molecular basis of pectin has helped us to understand some aspects of this complex polysaccharide. Extensive studies must be carried out to find out more about the biological pathways to devise various efficient means of pectin extraction that are scalable and can be commercialized. The large variety of applications as well as the increasing number of studies on pectin suggests that the potential of pectin as novel and versatile biomaterial will be even more significant in the future. As the research and development continues in pectin-based products, we expect to see many innovative and exciting applications.

Acknowledgements

The authors would like to thank Fr. Jobi, Head of the Department of Life Sciences, for providing laboratory facilities and supporting this work.

Conflict of interest

The authors would like to declare that there was no conflict of interest in this work.

Author details

Erumalla Venkatanagaraju*, N. Bharathi, Rachiraju Hema Sindhuja, Rajshree Roy Chowdhury and Yarram Sreelekha
Department of Life Sciences, CHRIST (Deemed to be University), Bangalore, Karnataka, India

*Address all correspondence to: venkatanagarajue@gmail.com

References

[1] Bonnin E, Garnier C, Ralet MC. Pectin-modifying enzymes and pectin-derived materials: Applications and impacts. Applied Microbiology and Biotechnology. 2014;**98**(2):519-532

[2] Willats WG, Knox JP, Mikkelsen JD. Pectin: New insights into an old polymer are starting to gel. Trends in Food Science and Technology. 2006;**17**(3):97-104

[3] Munarin F, Tanzi MC, Petrini P. Advances in biomedical applications of pectin gels. International Journal of Biological Macromolecules. 2012;**51**(4):681-689

[4] Voragen AGJ, Coenen GJ, Verhoef RP, Schols HA. Pectin, a versatile polysaccharide present in plant cell walls. Structural Chemistry. 2009;**20**(2):263-275

[5] Liu L. Pectin composite matrices for biomedical applications. Biomaterials. 2004;**25**(16):3201-3210

[6] Thakur BR, Singh RK, Handa AK, Rao MA. Chemistry and uses of pectin—A review. Critical Reviews in Food Science and Nutrition. 1997;**37**(1):47-73

[7] Willats WGT, McCartney L, Mackie W, Knox JP. In: Carpita NC, Campbell M, Tierney M, editors. Pectin: Cell Biology and Prospects for Functional Analysis in Plant Cell Walls. Dordrecht: Springer Netherlands; 2001. pp. 9-27

[8] Ciriminna R, Chavarria N. Pectin: A new perspective from the biorefinery standpoint. Biofuels, Bioproducts and Biorefining. 2015;**9**(4):368-377

[9] Jarvis MC. Structure and properties of pectin gels in plant cell walls. Plant, Cell & Environment. 1984;**7**(3):153-164

[10] Valdes A, Burgos N, Jimenez A, Garrigos M. Natural pectin polysaccharides as edible coatings. Coatings. 2015;**5**(4):865-886

[11] Mishra RK, Banthia AK, Majeed ABA. Pectin based formulations for biomedical applications: A review. Asian Journal of Pharmaceutical and Clinical Research. 2012;**5**(4):1-6

[12] Pacheco MT, Villamiel M, Moreno R, Moreno FJ. Structural and rheological properties of pectins extracted from industrial sugar beet by-products. Molecules. 2019;**24**(3):392

[13] Mohnen D. Biosynthesis of pectins and galactomannans. Comprehensive Natural Products Chemistry. 1999;**3**(3):497-527

[14] Mohnen D. Pectin structure and biosynthesis. Current Opinion in Plant Biology. 2008;**11**(3):266-277

[15] McNeil M, Darvill AG, Albersheim P. The structural polymers of the primary cell walls of dicots. In: Herz W, Grisebach H, Kirby GW, editors. Progress in the Chemistry of Organic Natural Products. Vol 37. New York: Springer-Verlag; 1979. pp. 191-249

[16] Schols H, Voragen A. Complex pectins: Structure elucidation using enzymes. Pectins and Pectinases. 1996;**14**:3-19

[17] Oneill M, Albersheim P, Daevill A. The pectic polysaccharides of primary cell walls. Methods in Plant Biochemistry. 1990;**2**:415-441

[18] Harris PJ, Smith BG. Plant cell walls and cell-wall polysaccharides: Structures, properties and uses in food products. International Journal of Food Science and Technology. 2006;**41**(2):129-143

[19] McNeil M, Darvill AG, Albersheim P. Structure of plant cell walls: X. Rhamnogalacturonan I, a structurally

complex pectic polysaccharide in the walls of suspension-cultured sycamore cells. Plant Physiology. 1980;**66**(6):1128-1134

[20] Kravtchenko T, Arnould I, Voragen A, Pilnik W. Improvement of the selective depolymerization of pectic substances by chemical β-elimination in aqueous solution. Carbohydrate Polymers. 1992;**19**(4):237-242

[21] Ishii T, Matsunaga T. Pectic polysaccharide rhamnogalacturonan II is covalently linked to homogalacturonan. Phytochemistry. 2001;**57**(6):969-974

[22] Nakamura A, Furuta H, Maeda H. Structural studies by stepwise enzymatic degradation of the main backbone of soybean soluble polysaccharides consisting of galacturonan and rhamnogalacturonan. Bioscience, Biotechnology, and Biochemistry. 2002;**66**(6):1301-1313

[23] Ridley BL, Neill MA, Mohnen D. Pectins: Structure, biosynthesis, and oligogalacturonide-related signaling. Phytochemistry. 2001;**57**(6):929-967

[24] Bonner J. The chemistry and physiology of the pectins. The Botanical Review. 1936;**2**(10):475-497

[25] Norziah MH, Fang EO, Karim AA. Extraction and Characterization of Pectin from PomeloFruit Peels. In: Williams PA, editor. Gums and Stabilisers for the Food Industry. Vol 10. Cambridge, UK: The Royal Society of Chemistry; 2000. pp. 26-36

[26] Mudgil D, Barak S. Composition, properties and health benefits of indigestible carbohydrate polymers as dietary fiber: A review. International Journal of Biological Macromolecules. 2013;**61**:1-6

[27] Canteri-schemin MH, Cristina H, Fertonani R, Waszczynskyj N. Extraction of pectin from apple pomace. Brazilian Archives of Biology and Technology. 2005;**48**:259-266

[28] Zafari F, Khodaiyan F, Kiani H, Hosseini SS. Pectin from carrot pomace: Extraction optimization and physicochemical properties of pectin from melon peel. International Journal of Biological Macromolecules. 2017;**98**:709-716

[29] Huang X, Li D, Wang L. Characterization of pectin extracted from sugar beet pulp under different drying conditions. Journal of Food Engineering. 2017;**211**:1-6

[30] Levigne S, Ralet MC, Thibault JF. Characterisation of pectins extracted from fresh sugar beet under different conditions using an experimental design. Carbohydrate Polymers. 2002;**49**(2):145-153

[31] Shi XQ, Chang KC, Schwarz JG, Wiesenborn DP, Shih MC. Optimizing pectin extraction from sunflower heads by alkaline washing. Bioresource Technology. 1997;**58**:291-297

[32] Koubala BB, Christiaens S, Kansci G, Van Loey AM, Hendrickx ME. Isolation and structural characterisation of papaya peel pectin. Food Research International. 2014;**55**:215-221

[33] Yeoh S, Shi J, Langrish T. Comparisons between different techniques for water-based extraction of pectin from orange peels. Desalination. 2008;**218**(1):229-237

[34] Yapo BM, Koffi KL. Extraction and characterization of gelling and emulsifying pectin fractions from cacao pod husk. Journal of Food and Nutrition Research. 2013;**1**(4):46-51

[35] Gan CY, Latiff AA. Extraction of antioxidant pectic-polysaccharide from mangosteen (*Garcinia mangostana*) rind: Optimization using response surface

methodology. Carbohydrate Polymers. 2011;**83**(2):600-607

[36] Wai WW, Alkarkhi AFM, Easa AM. Effect of extraction conditions on yield and degree of esterification of durian rind pectin: An experimental design. Food and Bioproducts Processing. 2010;**88**(2):209-214

[37] Bagde PP, Dhenge S, Bhivgade S. Extraction of pectin from orange peel and lemon peel. International Journal of Engineering Technology Science and Research. 2017;**4**(3):1-7

[38] Zaidel DNA, Rashida J, Hamidona NH, Salleha L, Kassim ASM. Ultrasound assisted extraction of pectin from dragon fruit peels. Chemical engineering transactions. 2017;**56**:805-810

[39] Bhavya DK, Suraksha R. Value added products from agriculture: Extraction of pectin from agro waste product *Musa Acuminata* and citrus fruit. Food and Industrial Microbiology. 2015;**3**:6

[40] Edima H, Biloa D, Enama T, Abossolo S, Mbofung C. Optimization of the extraction of pectin from *Cucumis melo*. British Journal of Applied Science & Technology. 2014;**4**(35):4860-4877

[41] Sarah M, Hanum F, Rizky M, Hisham MF. Microwave-assisted extraction of pectin from cocoa peel. Conference Series: Earth and Environmental Science. 2018;**122**:012079

[42] Fishman ML, Chau HK, Hoagland PD, Hotchkiss AT. Microwave-assisted extraction of lime pectin. Food Hydrocolloids. 2006;**20**(8):1170-1177

[43] Campbell M. Extraction of pectin from watermelon rind [thesis]. 2006. pp. 1-96

[44] Kanse NG, Chirag S, Vishal S. Effects of operating parameters on extraction of pectin from orange peel's. International Journal of Engineering Technology Science and Research. 2018;**5**(4):981-987

[45] Hamidon NH, Zaidel DNA. Effect of extraction conditions on pectin yield extracted from sweet potato peels residues using hydrochloric acid. Chemical Engineering Transactions. 2017;**56**:979-984

[46] Yousuf O, Singh A, Shahi NC, Kumar A, Verma AK. Ultrasound assisted extraction of pectin from orange peel. Bulletin of Environment, Pharmacology and Life Sciences. 2018;**7**(12):48-54

[47] Kanmani P. Extraction and analysis of pectin from citrus peels: Augmenting the yield from citrus limon using statistical experimental design. Iranica Journal of Energy and Environment. 2014;**5**(3):303-312

[48] Shaha RK. Optimized extraction condition and characterization of pectin from kaffir lime (*Citrus hystrix*). Research Journal of Agriculture and Forestry Sciences. 2013;**1**(2):1-11

[49] Sathish S, Gowthaman KA, Augustus H. Utilization of *Punica granatum* peels for the extraction of pectin. Research Journal of Pharmacy and Technology. 2018;**11**(2):613

[50] Pandharipande S, Makode H. Separation of oil and pectin from orange peel and study of effect of pH of extracting medium on the yield of pectin. Journal of Engineering Research and Studies. 2012;**3**(2):6-9

[51] Aina VO. Extraction and characterization of pectin from peels of lemon (*Citrus limon*), grape fruit (*Citrus paradisi*) and sweet orange (*Citrus sinensis*). British Journal of Pharmacology and Toxicology. 2012;**3**(6):259-262

[52] Liu Y, Shi J, Langrish T. Water-based extraction of pectin from flavedo and albedo of orange peels.

Chemical Engineering Journal. 2006;**120**(3):203-209

[53] Grassino AN, Brncic M, Vikic D, Roca S, Dent M, Brncic CR. Ultrasound assisted extraction and characterization of pectin from tomato waste. Food Chemistry. 2016;**198**:93-100

[54] Hamed A, Elkhedir A, Mustafa S. Effect of Soxhlet method extraction on characterization of pectin of pumpkin peels. Journal of Experimental Food Chemistry. 2017;**3**(1):1-3

[55] Azad A, Ali M, Akter MS, Rahman MJ, Ahmed M. Isolation and characterization of pectin extracted from lemon pomace during ripening. Journal of Food and Nutrition Sciences. 2014;**2**(2):30-35

[56] Gomez B, Gullon B, Yanez R, Parajo JC, Alonso JL. Pectic oligosacharides from lemon peel wastes: Production, purification, and chemical characterization. Journal of Agricultural and Food Chemistry. 2013;**61**(42):10043-10053

[57] Gama B, Silva CDF, Da Silva LO, Abud A. Extraction and characterization of pectin from citric waste. Chemical Engineering Transactions. 2015;**44**:259-264

[58] Virk B, Sogi D. Extraction and characterization of pectin from apple (*Malus pumila Cv amri*) peel waste. International Journal of Food Properties. 2004;**7**(3):693-703

[59] Ogunka-Nnoka CU, Atinlikou MF. Extraction and characterization of pectin from some selected non-citrus agricultural food wastes. Journal of Chemical and Pharmaceutical Research. 2016;**8**:283-290

[60] Girma E, Worku T. Extraction and characterization of pectin from selected fruit peel waste. International Journal of Scientific and Research Publications. 2016;**6**(2):447-454

[61] Castillo-Israel K, Baguio S, Diasanta M, Lizardo R, Dizon E, Mejico M. Extraction and characterization of pectin from Saba banana [*Musa'saba* (*Musa acuminata × Musa balbisiana*)] peel wastes: A preliminary study. International Food Research Journal. 2015;**22**(1):202-207

[62] Liew SQ, Chin NL, Yusof YA. Extraction and characterization of pectin from passion fruit peels. Agriculture and Agricultural Science Procedia. 2014;**2**:231-236

[63] Panchami P, Gunasekaran S. Extraction and characterization of pectin from fruit waste. International Journal of Current Microbiology and Applied Sciences. 2017;**6**(8):943-948

[64] Sudhakar D, Maini S. Isolation and characterization of mango peel pectins. Journal of Food Processing and Preservation. 2000;**24**(3):209-227

[65] Sundarraj AA, Vasudevan RT, Sriramulu G. Optimized extraction and characterization of pectin from jackfruit (*Artocarpus integer*) wastes using response surface methodology. International Journal of Biological Macromolecules. 2018;**106**:698-703

[66] Tamaki Y, Konishi T, Fukuta M, Tako M. Isolation and structural characterisation of pectin from endocarp of *Citrus depressa*. Food Chemistry. 2008;**107**(1):352-361

[67] Tiwari AK, Saha SN, Yadav VP, Upadhyay UK, Katiyar D, Mishra T. Extraction and characterization of pectin from orange peels. International Journal of Biotechnology and Biochemistry. 2017;**13**(1):39-47

[68] Madhav A, Pushpalatha P. Characterization of pectin extracted from different fruit wastes. Journal of Tropical Agriculture. 2006;**40**:53-55

[69] Khamsucharit P, Laohaphatanalert K, Gavinlertvatana P, Sriroth K,

Sangseethong K. Characterization of pectin extracted from banana peels of different varieties. Food Science and Biotechnology. 2018;**27**(3):623-629

[70] Noreen A. Pectins functionalized biomaterials; a new viable approach for biomedical applications: A review. International Journal of Biological Macromolecules. 2017;**101**:254-272

[71] Katav T, Liu L, Traitel T, Goldbart R, Wolfson M, Kost J. Modified pectin-based carrier for gene delivery: Cellular barriers in gene delivery course. Journal of Controlled Release. 2008;**130**(2):183-191

[72] Sriamornsak P. Application of pectin in oral drug delivery. Expert Opinion on Drug Delivery. 2011;**8**(8):1009-1023

[73] Ninan N, Muthiah M, Park IK, Elain A, Thomas S, Grohens Y. Pectin/carboxymethyl cellulose/micro fibrillated cellulose composite scaffolds for tissue engineering. Carbohydrate Polymers. 2013;**98**(1):877-885

[74] Swenson H, Miers J, Schultz T, Owens H. Pectinate and pectate coatings. 2. Application to nuts and fruit products. Food Technology. 1953;**7**(6):232-235

[75] Colin D. Industrial pectins: Sources, production and applications. Carbohydrate Polymers. 1990;**12**:79-99

[76] El-Shamei Z, El-Zoghbi M. Producing of natural clouding agents from orange and lemon peels. Food/Nahrung (Food Nahrung). 1994;**38**(2):158-166

[77] Danielle BP, Alexandre CM, Eleni G. Pectin and pectinases: Production, characterization and industrial application of microbial pectinolytic enzymes. 2009;**3**:9-18

[78] Liu L, Finkenstadt V, Liu C, Jin T, Fishman M, Hicks K. Preparation of poly (lactic acid) and pectin composite films intended for applications in antimicrobial packaging. Journal of Applied Polymer Science. 2007;**106**(2):801-810

[79] Fishman M, Coffin D, Konstance R, Onwulata C. Extrusion of pectin/starch blends plasticized with glycerol. Carbohydrate Polymers. 2000;**41**(4):317-325

[80] Liu L, Liu CK, Fishman ML, Hicks KB. Composite films from pectin and fish skin gelatin or soybean flour protein. Journal of Agricultural and Food Chemistry. 2007;**55**(6):2349-2355

[81] Mishra RK, Majeed ABA, Banthia AK. Development and characterization of pectin/gelatin hydrogel membranes for wound dressing. International Journal of Plastics Technology. 2011;**15**(1):82-95

[82] Sriamornsak P, Wattanakorn N, Takeuchi H. Study on the mucoadhesion mechanism of pectin by atomic force microscopy and mucin-particle method. Carbohydrate Polymers. 2010;**79**(1):54-59

[83] Fisher A, Watling M, Smith A, Knight A. Pharmacokinetics and relative bioavailability of fentanyl pectin nasal spray 100-800 μg in healthy volunteers. International Journal of Clinical Pharmacology and Therapeutics. 2010;**48**(12): 860-867

[84] Portenoy RK, Burton AW, Gabrail N, Taylor D. A multicenter, placebo-controlled, double-blind, multiple-crossover study of fentanyl pectin nasal spray (FPNS) in the treatment of breakthrough cancer pain. Pain. 2010;**151**(3):617-624

[85] Illum L. Nasal drug delivery composition containing nicotine. Patent; 1999. 1-14

[86] Paharia A, Yadav AK, Rai G, Jain SK, Pancholi SS, Agrawal

GP. Eudragit-coated pectin microspheres of 5-fluorouracil for colon targeting. AAPS PharmSciTech. 2007;**8**(1):87-93

[87] Glinsky VV, Raz A. Modified citrus pectin anti-metastatic properties: One bullet, multiple targets. Carbohydrate Research. 2009;**344**(14):1788-1791

[88] Pienta KJ. Inhibition of spontaneous metastasis in a rat prostate cancer model by oral administration of modified citrus pectin. Journal of the National Cancer Institute. 1995;**87**(5):348-353

[89] Kuwabara I, Liu FT. Galectin-3 promotes adhesion of human neutrophils to laminin. The Journal of Immunology. 1996;**156**(10):3939-3944

[90] Platt D, Raz A. Modulation of the lung colonization of B16-F1 melanoma cells by citrus pectin. Journal of the National Cancer Institute. 1992;**84**(6):438-442

[91] Ledley FD. Nonviral gene therapy: The promise of genes as pharmaceutical products. Human Gene Therapy. 1995;**6**(9):1129-1144

[92] Monsigny M, Midoux P, Mayer R, Roche AC. Glycotargeting: Influence of the sugar moiety on both the uptake and the intracellular trafficking of nucleic acid carried by glycosylated polymers. Bioscience Reports. 1999;**19**(2):125-132

[93] Brown MD, Schatzlein AG, Uchegbu IF. Gene delivery with synthetic (non viral) carriers. International Journal of Pharmaceutics. 2001;**229**(1):1-21

[94] Asnaghi MA. Trends in biomedical engineering: Focus on regenerative medicine. Journal of Applied Biomaterials & Biomechanics. 2011;**9**(2):73-86

[95] Wightman L. Different behavior of branched and linear polyethylenimine for gene delivery in vitro and in vivo. The Journal of Gene Medicine. 2001;**3**(4):362-372

[96] Opanasopit P, Apirakaramwong A, Ngawhirunpat T, Rojanarata T, Ruktanonchai U. Development and characterization of pectinate micro/nanoparticles for gene delivery. AAPS PharmSciTech. 2008;**9**(1):67-74

[97] Burapapadh K, Takeuchi H, Sriamornsak P. Development of pectin nanoparticles through mechanical homogenization for dissolution enhancement of itraconazole. Asian Journal of Pharmaceutical Sciences. 2016;**11**(3):365-375

[98] Coimbra P. Preparation and chemical and biological characterization of a pectin/chitosan polyelectrolyte complex scaffold for possible bone tissue engineering applications. International Journal of Biological Macromolecules. 2011;**48**(1):112-118

[99] Munarin F. Pectin-based injectable biomaterials for bone tissue engineering. Biomacromolecules. 2011;**12**(3):568-577

[100] Xiao C, Anderson CT. Roles of pectin in biomass yield and processing for biofuels. Frontiers in Plant Science. 2013;**4**(67):1-7

[101] Ludwig A. The use of mucoadhesive polymers in ocular drug delivery. Advanced Drug Delivery Reviews. 2005;**57**(11):1595-1639

Section 2

Pectin Applications

Chapter 3

Pectin - Extraction, Purification, Characterization and Applications

Virginia Rodríguez Robledo and Lucía Isabel Castro Vázquez

Abstract

Fruits, vegetables, and other plant-based foods are particularly important as they are source of dietary carbohydrates, and therefore much of the energy in the adult diet. Plant food also contains a wide range of dietary components rich in bioactive phytochemicals and is essential to the human body that may provide desirable health benefits beyond basic nutrition. Pectin is one of the nonstarch polysaccharides (NSPs), which constitutes the major fraction of the plant cell wall in association and/or substituted with other polysaccharides, and they cover a great variety of biological functions and chemical structures. Generally, pectin is isolated from by-products of agro-foods using extraction technologies with the emergence of novel and effective techniques that inclined toward a cleaner process. Pectin is widely used both in food sector (as gelling, thickening, and stabilizer agent) and in pharmaceutical industries (bioactive components) including biomedical application (drug delivery, tissue engineering, and wound healing) as innovative applications.

Keywords: new sources, pectin isolations, innovative application, food sector, tissue engineering, drug delivery

1. Introduction

Pectins represent a group of structurally heteropolysaccharides, composed mainly by covalently α-1,4-linked D-galacturonic acid (GalA) units, found in primary cell walls and middle lamella of higher plants [1, 2]. Pectin contributes to the firmness and structure of plant tissue, being involved in the intercellular adhesion and mechanical resistance of the cell wall. They also have an important role in the development of plant cells providing turgidity and resistance [3] **Figure 1**. These polysaccharides have been used in the food and beverage industries for many years. The principal applications of pectin are as a gelling agent, stabilizer, emulsifier, and thickening agent [4–6].

In the food sector, this traditional usage is being complemented by the emerging utilization of pectin as a fat replacer and health-promoting functional ingredient [4, 7, 8]. Pectin also provides an important source of dietary fiber that has been identified as emerging prebiotic with improved therapeutic properties for gut microbiota modulation [9–11].

Pectin has also been used for medicine and pharmaceutical purposes as a carrier for controlled drugs or bioactive release [12], for example, in drug delivery [13, 14].

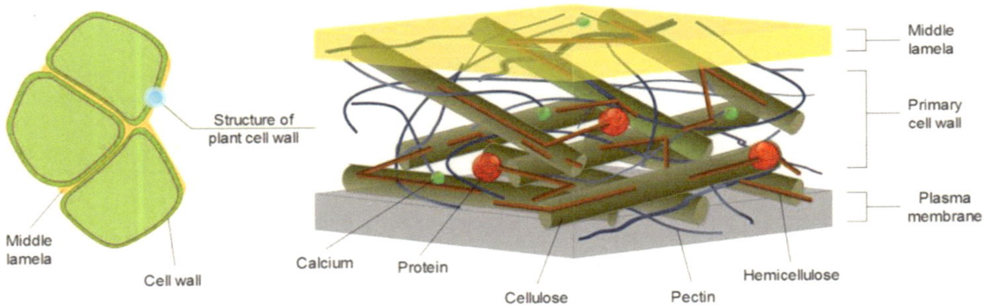

Figure 1.
Structure of primary plant cell wall (Copyright Figure 1, [56]).

2. Pectin chemical structure

Nowadays, there are a consensual classification of pectin based on three main structural domains: homogalacturonan (HGA), alternating with two types of highly branched rhamnogalacturonan regions designated as RG-I and RG-II [2]. Other structural classes of pectic polysaccharides also comprise xylogalacturonan, apiogalacturonan, arabinan, galactan, and arabinogalactan I [15, 16]. The pectin structure largely governs its physicochemical properties and its applications for several purposes. In this context, it is important to note that the carboxylic groups or hydroxyls may be methyl-esterified and/or O-acetyl-esterified [16]. The O-acetyl-esterification occurs predominantly at the O-3 position and occasionally at the O-2 position. However, not only these chemical variables are important but also the composition of neutral sugars, the degree of branching, and the degree of polymerization (molar weight), which can also modify the structure of this complex polysaccharide [17, 18]. The pectin gelation process is strongly influenced by the pectin esterification degree (DM). So, depending on the DM of pectin (defined as percentage of carboxyl groups esterified) [19], two different procedures of pectin gelation can be distinguished. High-methoxylated pectins (DM > 50%) are pectin formed in the presence of cosolutes as sucrose, at a concentration > 55%, and under pH < 3.5 [20]. Low-methoxylated pectins (DM < 50%) are gels created with the presence of divalent cations, mainly Ca^{2+}. Gelation process is due to the formation of junction zones between HGA regions of different pectin chains through calcium bridges between dissociated carboxyl groups [21].

High and low methoxyl pectins have diverse physicochemical properties and therefore have varied applications. High methoxyl pectin can be used as a gelling agent, emulsifier, stabilizer, and thickener in the food industry for the production of jams and jellies. Low methoxyl pectin can be used like a fat replacer, ice cream, yoghurt, bakery glazing, emulsified meat products, and low calorie products [22]. The diversity of pectin structures influences the physicochemical properties and also in its different technological applications, biological activities, biofunctionality, and therapeutic properties.

2.1 Homogalacturonan (HGA)

HGA is a linear homopolymer of α-1,4-linked GalA and is known as the "smooth region." It is an abundant and widespread domain of pectin, accounting for approximately 60–65% of total pectin amount [23]. The carboxyl groups present at C-6 in the GalA units can be partially methyl-esterified. Around 70–80% of GalA units, methyl-esterified could also be O-acetyl-esterified at O-2 or O-3 positions, depending on its provenance [2]. The amount of GalA units present in a HGA chain is estimated to be around 100–200 units [24]. The smooth region can sometimes be joined by one or two α-1,2-linked L-rhamnopyranose units and most of the pectins have this structure. In addition, GalA units may be substituted at the C-2 or C-3 positions with residues of xylose or apiose, producing domains known as xylogalacturonan or apiogalacturonan, respectively [25].

Regarding other structural classes of pectin polysaccharides, xylogalacturonans (XGAs) are HGA substituted with β-linked-D-xylose-(1-3) at O-3 single unit side chains [26]. The degree of xylosidation can vary between 25 and 75% for instance in watermelon and apple, respectively [27]. Part of the GalA residues in XGA is methyl-esterified independently of the xylose substitutions [28].

2.2 Rhamnogalacturonan I (RG-I)

RG-I comprises a backbone of repeating disaccharide units of [-α-L-Rhap-1,4-α-D-GalpA-1,2-]$_n$ [29, 30]. The RG-I backbone may enclose up to 300 rhamnosyl and 300 galactosyluronic acid residues. The composition of neutral sugar side chains can be a glycosyl residue up to 50, resulting in a large and highly variable family of polysaccharides with a range of glycosidic linkages [23]. The highly branched nature of RG-I has led to the name "hairy region." The GalA residues of RG-I could be O-acetyl-esterified but they are not methyl-esterified [31]. In most cases, rhamnose residues are preferably substituted at the C-4 position with neutral sugar side chains, like arabinose and galactose forming arabinan, galactan, and arabinogalactans [2, 32].

Regarding to arabinogalactans, RG-I with galactose and arabinose side chains possess two structural different forms designed as arabinogalactans I (AGI) and arabinogalactans II (AGII). The AGI consist in a linear chain of 1,4-linked β-D-galactose, containing up to 25% α-L-arabinose residues 1,5-linked in short side chains, connected predominantly to O-4 of the rhamnosyl residues [1]. This type of arabinogalactan has been widely spread in citrus, apple, potatoes, soybean, lupin, cabbage, onion, tomato, and kiwi. It is know that HG and RGI are covalently linked and cannot be separated without a chain-cleavage using, for example, enzymes such as endopolygalacturonase or chromatographic fractionation methods [33]. AGII similarly contains chains of β-D-galactopyranosyl units, but glycosidic linkages occur at C-1, C-3, and C-6 in galactose molecules [34]. This type of arabinogalactan is highly ramified, and it may also include an L-arabinopyranosyl residue at the end of the chain of β-1,6-D-galactopyranosyl units.

2.3 Rhamnogalacturonan-II (RG-II)

RG-II is a homopolymer composed of a backbone that consists of 7–11 galacturonic acid residues branched with 4 side chains, preferably at C-2 and C-3, which may include apiose, 2-O-methyl-L-fucose, 2-O-methyl-D-xylose, 3-C-carboxy-5-deoxy-L-xylose (aceric acid), 3-deoxy-D-manno-octulosonic acid, and 3-deoxy-D-lyxoheptulosaric acid [15, 35]. The side chains of RGII consist of 12 different types of sugars with over 20 different linkages [2]. The RGII, the most structurally complex pectin domain, is highly conserved across many plant sources. In addition, RG-II has the ability to form borate esters dimers [36].

3. Food sources

Nowadays, fruit and some food by-products can be considered as raw materials to produce value-added products enriched with pectin. Pectin is not only a gelling agent but also a thickening, stabilizing, and emulsifying agent [37], and even it has been used as a fat replacer and health-promoting functional ingredient [4, 8]. Thus, pectin must be taken into account within the set of new opportunities for the development of innovative products.

3.1 Conventional sources

Historically, commercial pectin is mostly extracted from citrus peel (85.5%), followed by apple pomace (14.0%) and, to a smaller extent, of the sugar industry (0.5%) [7, 38–42]. In the cases of apple, citrus, and sugar beet, where the backbone of HG is formed by at least 72–100 D-galacturonic acid residues, the degree of methylation is one of the key parameters related to gelling aptitude. The gelation process is affected by factors including the number and distribution of free carboxyl groups and the molecular weight, but also by other factors as calcium concentration, pH, ionic strength, and temperature [1, 43].

Regarding to citrus fruit, the amount of pectin has been estimated to account for as much as 13.4–29.1% of the dry weight [44]. The yield of pectin recovery from lime, orange, lemon and, lime peels using enzymatic extraction showed no significant effect compared with the extraction yields obtained using acid isolation technique [45].

The second most used source of pectin is apple pulp and apple pomace containing among 4.2–19% of pectin rich in neutral sugars as arabinose, galactose, rhamnose, glucose, and xylose [42]. Apple pectin displays a more elastic-viscous gel; however, citrus pectin produces a more elastic-brittle gel [46, 47].

Sugar beet pulp is also considered as an outstanding low cost source for commercial pectin coming from the sugar industry activity [32]. The combined effect of reduced pH, increased temperature, extraction time, and solvent-solid relation of the extraction process have been related with higher recovery rates of sugar beet pectin ranged between 6.5 and 24.8% [48]. Sugar beet pectin are characterized by showing a diminished gelling properties that have been attributed to its higher acetyl group content, lower molecular weight, higher amount of proteins bound covalently in the lateral chains and greater neutral sugar level [49–51]. Nevertheless, sugar beet pectin processes remarkable emulsifying properties, greater to pectin extracted from other conventional sources, as consequence of the ferulic acid residues on the arabinan side chains of RG-I, which allow to create covalently cross-linked hydrogels increasing its industrial potential [52, 53].

3.2 Nonconventional sources

Currently, other vegetable wastes constitute a new source of pectin with interesting food applications as valorizing new products and as functional ingredient with health-promoting benefits [54–56]. Nowadays, there are several unconventional feasible sources of pectin coming from food, vegetable residues, and plant species, which have different pectin contents and physicochemical properties. In this sense, sunflower head residues possess a very interesting gelling properties based on its high molecular weight and high GalA content. This kind of pectin contains around 3.5–5.0% water soluble high-methoxyl pectin and between 12 and 14% of insoluble low-methoxyl pectin [57].

Pectin from olive pomace, a semi-solid by-product from the olive oil industry has low molecular weight and high content of neutral sugars (as beet pectin) and a high percentage of acetylation, which facilitate the formation of gel by ionic interactions with calcium [58, 59]. Among 5–8% of pectin from olive pomace have

Extraction technique	Plant source	Operating conditions	Pectin content (%)	Effect on yield and quality	Ref.
Solvent extraction (SE)	Cocoa husks	Solvent: water, citric acid or hydrochloric acid; pH: 2.5 or 4.0; temperature: 50 or 95°C; time: 1.5 or 3.0 h	3.38–12.60	The highest pectin yield (7.62%) was obtained using citric acid (pH 2.5, 95°C, 3.0 h), while the highest uronic acid content in pectin (65.20%) resulted by using water (95°C, 3.0 h). Extraction with citric acid produced pectin with a wider DM range.	[86]
	Sour orange peel	Solvent: water; liquid/solid ratio: 20:1, 30:1, or 40:1 (v/w); temperature: 75, 85, or 95°C; time: 30, 60, or 90 min	17.95	The yield of pectin extracted at the optimal condition (95°C, 90 min, and liquid/solid ratio of 25:1) was 18.35%. GalA content and DE of the extracted pectin ranged from 57 to 83%, and 17–30.5%, respectively.	[87]
	Durian rinds	Solvent: water; solid/liquid ratio: 1:5–1:15 (g/mL); pH: 2–3; temperature: 75–95°C; time: 20–60 min	9.10	Under optimal conditions (solid/liquid ratio of 1:10 g/mL, pH of 2.8, 43 min, 86°C)	[88]
	Lime peel	Solvent: water: nitric acid; pH: 1.5, 2.3, or 3.1; temperature: 60, 70, or 80°C	15.7	Higher pectin solution concentrations were obtained at the lower pH values. Increased temperature and especially acidity caused a faster decrease of DE, effect that was particularly significant during extractions at pH = 1.5.	[89]
	Apple pomace	Enzymes: endo-xylanase and endo-cellulaseb; solid/liquid ratio: 1 g/15 mL; enzyme dose: 50 U/g; pH: 5.0; temperature: 40°C; time: 10 h	6.85	Treatment with endo-xylanase resulted in the highest pectin yield (19.8%) and very high DM (73.4%). Pectin extracted by endo-cellulase treatment was characterized by the high GalA content (70.5%). Simultaneous use of both enzymatic preparations resulted in a 10.2% extraction yield and a pectin rich in galacturonic acid (74.7%).	[90]
	Rapeseed cake	Commercial enzymes: celluclast and alcalase; enzyme/rapeseed cake ratio: 1:50–1:65 (v/w); celluclast/alcalase ratio: 0:5, 1:4, 2:3, 3:2, 4:1, or 5:0 (v/v); time: 90, 180, 270, 360, or 450 min	6.85	Pectin yield (6.85%) without significant loss of GalA was a 1:50 enzyme/RSC ratio with a celluclast/alcalase ratio of 1:4 for a 270 min hydrolysis time. Enzymes indicated different functions: alcalase led to the destruction of protein-carbohydrate complexes, while celluclast slightly cleaved some linkages of carbohydrate.	[91]
	Lime peel	Enzymes: Laminex C2K, Multifect B, GC220, and GC880; pH: 3.5–6.5; temperature: 40–70°C	23	Laminex C2K preparation proved to the most effective, as in optimum conditions (4 h treatment at pH 3.5, 50°C) gave a high yield (23%) and a pectin with good composition and properties (gelling, stabilization).	[92]

Extraction technique	Plant source	Operating conditions	Pectin content (%)	Effect on yield and quality	Ref.
Ultrasound-assisted extraction (UAE)	Passion fruit peel	Extraction solvent: 1.0 mol/L HNO3; pH 2.0, peel/solvent ratio of 1:30 (g/mL); temperature: 45–85°C; power intensity: 132.8–664.0 W/cm^2; time and frequency (constant): 10 min, 20 kHz	10.0–30.30	The highest pectin yield (12.67%) was obtained at 85°C and a power intensity of 664 W/cm^2. Despite the fact that pectin isolation reached the highest level, the isolate did not display the best composition, as the GalA content and DE showed minimum values.	[63]
	Mango peel	Extraction solvent: citric acid, pH 2.5, peel-solvent ratio of 1:40 (g/mL), time and frequency (constant): 15 min, 20 kHz; temperature: 20 or 85°C	9.20–31.80	Extraction yield of pectins varied greatly with the increase in temperature, from 2.09% (at 20°C) to 17.15% (at 85°C). A significant influence of temperature was also observed for GalA content (increase from 29.35 to 53.35%) and molecular weight (increase from 378.4 to 2320 kDa).	[93, 94]
	Jack fruit peel	Extraction solvent: distilled water); liquid-solid ratio: 10:1–20:1 (mL/g) pH: 1–2; sonication time: 15–30 min; extraction temperature: 50–70°C	8.94–14.5	Optimal conditions for the extraction were: liquid-solid ratio of 15:1 mL/g, pH of 1.6, sonication time of 24 min, and temperature of 60°C.	[95]
	Grapefruit peel	Emitter surface: 13 mm or 25 mm; power density: 0.20–0.53 W/mL; duty cycle: 33–80%; temperature: 30–80°C; solid-liquid ratio: 1/30–1/70 (g/mL); sonication time: 10–60 min	21.5	Heating significantly improved the extractability and extraction rate of pectin, leading to higher yield (26.74%) in shorter extraction time (52 min). The optimized parameters were: ultrasound power density 0.40 W/mL, duty cycle 50%, temperature 60°C, S/L 1/50 g/mL.	[8]
Subcritical water extraction (SWE)	Apple pomace and citrus peel	Solid to liquid ratio of 1:30 g/mL, extraction time of 5 min (constant); extraction temperature: 130, 150, or 170°C for apple pomace; 100, 120, or 140°C for citrus peel	14.0–21.9	The highest yield (21.95%) of citrus pectin (68.88% GalA content) was obtained at 120°C, and the highest yield of apple pectin (16.68%) was gained at 150°C.	[96]
	Sugar beet pulp	Temperature: 110, 120, or 130°C; extraction time: 20, 30, or 40 min; liquid/solid ratio: 30, 40, or 50 (w/w); extraction pressure: 8, 10, or 12 MPa	6.5–24.8	Increase in all parameters enhanced the extraction up to a certain level past which a decrease of pectin recovery was recorded. The optimum extraction conditions to obtain a maximum yield of 24.63% (with 59.12% GalA and 21.66% arabinose content) were as follows: L/S ratio of 44.03, 120.72°C extraction temperature, extraction time of 30.49 min and extraction pressure of 10.70 MPa.	[48, 97]

Extraction technique	Plant source	Operating conditions	Pectin content (%)	Effect on yield and quality	Ref.
Microwave-assisted extraction (MAE)	Waste papaya peel	Microwave power: 320, 480, or 640 W pH: 1, 2, or 3; time: 20, 100, or 180 s; solid-liquid ratio: 1:5, 1:15, or 1:25 g/mL	11.11–49.83	All the process variables had significant effect on pectin yield. The optimal conditions for reaching a maximum pectin yield (25.41%) were: microwave power of 512 W, pH of 1.8, time of 140 s, and solid-liquid ratio of 1:15 g/mL.	[98, 99]
	Waste mango peel	Microwave power: 160, 320, or 480 W; pH: 2, 3, or 4; time: 60, 120, or 180 s; solid-liquid ratio: 1:10, 1:20, or 1:30 g/mL	9.20–31.80	For all process parameters, a similar influence was observed: the increase in their level was positively correlated with the pectin yield up to a certain point, beyond which their effect on pectin extraction was negative. The maximum pectin yield (28.86%) could be obtained at a microwave power of 413 W, pH of 2.7, time of 134 s, and solid-liquid ratio of 1:18 g/mL	[88]
	Tangerine peels	Microwave power: 600, 700, or 800 W; temperature: 40, 50, or 60°C; time: 30, 40 or 50 s	19.9	The optimal extraction parameters were: microwave power 704 W, 52.2°C, extraction temperature, and extraction time of 41.8 min	[47]
	Opuntia ficus-indica cladodes	Microwave power: 200, 400, or 600 W pH: 1.5, 2.25, or 3; time: 1, 2, or 3 min; solid-liquid ratio: 2:20, 2:35, or 2:50 g/mL	12.56	The optimum conditions to obtain a maximum pectin recovery of 12.56% were 2.16 min, pH 2.26, 517 W microwave power and 2 g/30.66 mL of solid-liquid ratio. FTIR analysis indicated a GalA content of 34.4% and no alterations in the chemical structure of pectin following microwave treatment.	[100]

Extraction techniques and their effects on pectin yield. Adapted from [56].

Table 1.
Main sources of commercial pectin.

suitable emulsifying properties for commercial uses and also possess antioxidant and functional properties for its high dietary fiber levels. Carrot pomace and carrot peel, resulting from juice production, are vegetable wastes that possess approximately 22–25% of the total dietary fiber (29.6%) [60]. Pectin from carrot peels displays a very high level of linearity, a lower DM, and a better solubility than pectin extracted from carrot pomace [54] what makes it especially interesting for gelling mechanisms.

Pumpkin pulp and peel showed 7.46% of pectin under extraction at pH 4 with EDTA as chelating agent [61]. Moreover, the pumpkin cake is suggested to have a positive effect on gut bacteria [62]. In the same way, pectin from banana peels comprises around 1/4 of total fruit weight and contains a low amount of water-soluble pectin for gelling, thickening, or stabilizing purposes [63].

Watermelon rinds, which account for approximately one-third of total fruit mass and are usually discarded, were proposed as another possible source of pectin. The pectin content of wet watermelon rinds has displayed 19–21% (w/w)

[64] revealing its potential for use in the food industry. Both fresh and lyophilized watermelon pectin have showed high degree of methyl esterification and low molar mass [65]. These properties reveal its suitable application as emulsifier, stabilizing agent, and thickening agents.

Tomato residues, coming from the canning industry, lead large quantities of tomato pomace and tomato peels that contain 7.55 and 22.6% of pectin, respectively, expressed on a dry weight [66]. The structure of pectin molecules isolated from unripe tomato consist in complex biopolymer consists of HGs held together by RG-I regions [67, 68].

Another unconventional pectin sources are summarized in **Table 1**.

4. Pectin isolation process

Since pectin is a native polysaccharide that is found in the cell wall and middle lamellae of many land-growing plants, especially fruits and vegetables, an extraction process for its isolation is mandatory. The innovative applications (food and nonfood) that we will discuss in the following pages begin definitely with pectin isolation from the plant, fruit, or vegetal material.

This entire pectin production process is well documented in the literature and includes a raw material pretreatment stage, an extraction operation, and a postextraction stage (**Figure 2**). Generally, pectin is isolated by using chemical extraction process from by-products of agro-foods also called "Conventional Method." However, these methods require the disposal of, for example, acidic waste water that can cause serious environmental problems [69]. Currently, in connection with the emerging concept of "Green Chemistry," several novel or nonconventional extraction methods and more effective techniques have been developed for pectin extraction (**Table 1**).

4.1 Chemical extraction methods

Chemical extraction using acids are the most commonly employed methods in the production of pectin, and it has an impact on content and composition both of pectin monomeric as xylose, arabinose, and galacturonic acid and physicochemical properties (emulsifying activity and viscosity, among others) [70].

For example, the acid extraction from pumpkin pulp and peel using pH 4 with EDTA showed around 7.5% of pectin as chelating agent [61]. Higher yields of pectin were obtained (up to 200 mg g^{-1}) by combining the acid extraction with a microwave extraction, and those also displayed a large amount of phenolic compounds, proteins, and neutral sugars [18].

Other example for extraction yield of carrot pectin can range from 5.0 to 15.2% depending on parameters such as pH, temperature, heating time, sample provenance, and liquid/solid ratio [71].

The composition of the pectin from fabas and beans varied according to the extraction conditions: the neutral sugar galactose, arabinose, and rhamnose increased under slighter extraction conditions while glucose, mannose, and xylose sugars predominated under severe extraction. The maximum yield of extraction (15.75%) was recorded at pH 1.5 and at a temperature of 85°C for an 80-min extraction period and solid to liquid (1:25) ratio, while the highest degree of esterification (54.62%) occurred at pH 2.5 and at temperature of 90°C for a 60-min extraction period [72].

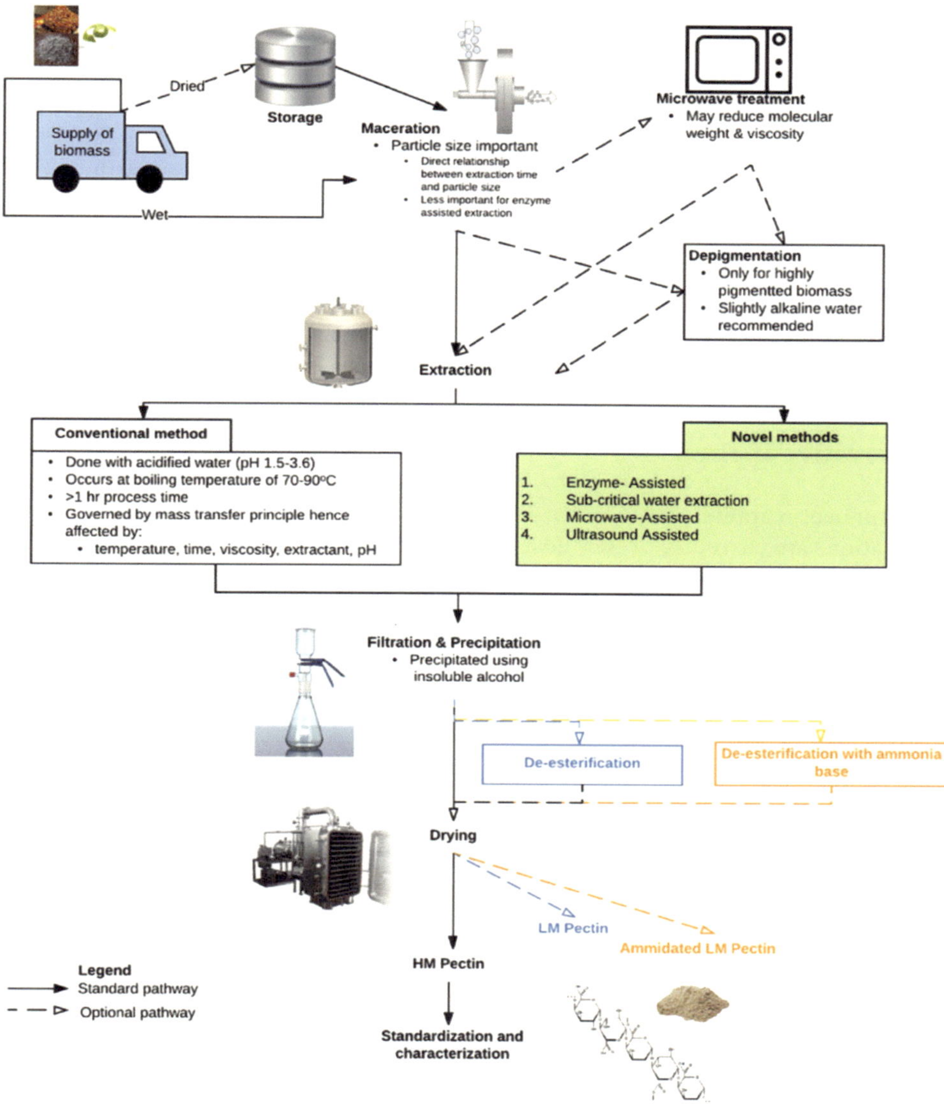

Figure 2.
The industrial pectin production process (Copyright Figure 1, [73]).

4.2 Novel technique extraction

Currently, novel and more effective techniques are inclined toward a cleaner (green or natural) extraction process. Among nonconventional extraction methods for pectin isolation are found (i) ultrasound-assisted extraction (UAE) (ii) subcritical water extraction (SWE), and (iii) microwave-assisted extraction (MAE). **Table 1** summarizes the main pectin extraction process including solvent extraction (SE).

Other relevant novel technique for pectin isolation is enzyme-assisted extraction (EAE), which among other benefits, the unavoidable presence of trace chemical solvents in products from solvent-based extraction processes [73]. The enzymes are characterized to catalyze reactions such as hydrolysis, with a high level of selectivity, which either reduces the amount of solvent/chemical needed or increase yield for the same amount of solvent [74], in a way that is not feasible with acid-based hydrolysis. A distinction is possible between two approaches of EAE of pectin: (i) using enzymes that degrade pectin and help isolate pectin fragments, for example, galacturonic acid and (ii) using enzymes capable of deconstructing plant cell wall and isolating pectin [75].

Concretely combined enzymatic and ultrasonic treatment has been evaluated for efficient extraction of pectin in comparison with other extraction methods. The sequential treatment proposed by Yang et al. with enzyme (Celluclast 1.5 L) followed by ultrasound of sisal waste attained a much higher pectin yield of 31.1% than the ultrasound followed by enzyme, 14.6%. The pectin yield attained with the combined enzymatic/ultrasonic extraction was also higher than enzymatic extraction (9.4%), ultrasonic extraction (11.9%), and acidic extraction (5.8%) [76].

5. Innovative applications

Main pectin applications have been focused to food sector, while nonfood applications are currently presented as innovative alternative. Nonfood application includes the use in medical and pharmaceutical industries, where the health promoting benefits and bioactivities of pectin has shown potential for biomedical applications including drug delivery, tissue engineering, and wound healing [77].

5.1 Food sector

In the food sector, pectin traditional usage as a gelling agent, thickening agent, and stabilizer is being complemented by the emerging utilization of pectin as a fat replacer and also as health-promoting functional ingredient [4, 7, 8, 56].

Apple pomace and citrus peel remain the main sources for the production of commercial pectins, although other sources are being considered (**Table 1**) to the rising demand and growing interest for the development of innovative products [20, 55, 56].

5.2 Nonfood pectin applications

Nonfood applications include the use of pectin in pharmaceutical industry, where pectin bioactivity has shown an outstanding potential for biomedical applications as bioactive components [12] and also includes drug and gene delivery [13, 14], tissue engineering [2], and wound healing patches [56].

Within biomedical application, drug delivery systems improve the conventional drug administration as they provide *in situ* controlled and sustained release of active biomolecules. For most drug delivery systems, natural polymers, where pectin is included, are employed as inert, biocompatible carriers. These compounds have interesting properties such as the mucoadhesiveness, the ease of dissolution in basic environments associated to its resistance to proteases and amylases, making pectin suitable to release drugs in the colon [77]. And the ability to form gels in acid environments, instead, enhances the contact time of drugs for gastric or ocular treatments [78, 79]. Furthermore, pectin has been found to recognize galectin molecules, which are involved in different stages of cancer pathologies, being particularly appealing to target tumor cells for chemiotherapic treatments [80].

Furthermore, pectin has been described as an emerging prebiotic with the ability to modulate the bacterial composition of the colon microbiota [81] being able to exert beneficial effects on health [70]. The prebiotic activity is one of the most outstanding health benefits. Pectin, indigestible food-ingredients, has

Type of dressing	Composition
Hydrocolloid sheets, uses: cavity or flat shallow wounds with low to medium exudates; absorbent; conformable; suitable for "problematic" areas: heel, elbow, sacrum	
CombiDERM® (ConvaTec Ltd.)	Cellulose, pectin, Salsorb90 (acrylic polymer)
DuoDERM® (ConvaTec Ltd.)	Carboxymethylcellulose, pectin, propylene glycol
Granuflex® (ConvaTec Ltd.)	Polyurethane foam sheet coated with pectin, gelatin and carboxymethylcellulose
Hydrocoll® (Hartmann)	Pectin, gelatin and carboxymethylcellulose
Hydrocolloid paste, uses: useful debriding agent, conformable, may be left in place for several days	
GranuGel® paste (ConvaTec Ltd.)	Pectin, carboxymethylcellulose, propylene glycol
CitruGel® (Advances medical)	Pectin, carboxymethylcellulose

Table 2.
Pectin-containing hydrocolloid wound dressing (Copyright Table 4, [77]).

considered as an emerging prebiotic that possess prebiotic effect into the colon microbiota [9–11]. The beneficial effects of the pectin-derived oligosaccharides are essentially explained by the positive effects caused into bacterial population [82]. The consumption of pectins generates increasing levels of acetate, propionate, and butyrate, derived from the intestinal fermentation of pectin. These short-chain fatty acids possess an outstanding role in the prevention and treatment of metabolic syndrome, diarrhea, and intestinal disorders mainly Crohn's disease and ulcerative colitis [83]. An example is the pumpkin cake, which is suggested to have a positive effect on gut bacteria [62].

Natural polymers have been employed as scaffolds to stimulate specific cell functions and to direct cell-cell interactions [77]. Few studies report the use of pectin for tissue regeneration; however, pectin hydrogels were found to have a great potential for bone tissue engineering applications, as they promote the nucleation of a mineral phase if immersed in adequate physiological solutions [84], with the formation of biomimetic constructs better mimicking the natural architecture of the bone [77].

Finally, other innovative applications in biomedical area are focused to wound healing patches. Hydrogel films on wounds or ulcers were aimed to prevent bacterial infection, support the autolytic debridement, and maintain a moist healing environment [77]. The natural properties of pectin impart several advantages to the wound dressings, such as hydrophilicity, which permits the removal of exudates, the retention of an acid environment, which may act as barrier against bacteria, and the ability for binding active molecules as drugs or growth factors to heal the wounds. As a wound heals, the cells around it are stimulated by growth factors to proliferate and grow into the wound [85].

A wide variety of hydrocolloid pectin-based wound dressings have been patented and are nowadays commercially available (**Table 2**).

Author details

Virginia Rodríguez Robledo* and Lucía Isabel Castro Vázquez
Faculty of Pharmacy, Department of Analytical Chemistry and Food Technology, University of Castilla-La Mancha, Albacete, Spain

*Address all correspondence to: virginia.rrobledo@uclm.es and luciai.castro@uclm.es

References

[1] Chan SY, Choo WS, Young DJ, Loh XJ. Pectin as a rheology modifier: Origin, structure, commercial production and rheology. Carbohydrate Polymers. 2017a;**161**:118-139

[2] Mohnen D. Pectin structure and biosynthesis. Current Opinion in Plant Biology. 2008;**11**(3):266-277

[3] Parre E, Geitmann A. Pectin and the role of the physical properties of the cell wall in pollen tube growth of solanum chacoense. Planta. 2005;**220**(4):582-592

[4] Min B, Bae IY, Lee HG, Yoo SH, Lee S. Utilization of pectin-enriched materials from apple pomace as a fat replacer in a model food system. Bioresource Technology. 2010;**101**(14):5414-5418

[5] Nakamura A, Furuta H, Maeda H, Takao T, Nagamatsu Y. Structural studies by stepwise enzymatic degradation of the main backbone of soybean soluble polysaccharides consisting of galacturonan and rhamnogalacturonan. Bioscience, Biotechnology, and Biochemistry. 2002;**66**(6):1301-1313

[6] Saha D, Bhattacharya S. Hydrocolloids as thickening and gelling agents in food: A critical review. Journal of Food Science and Technology. 2010;**47**(6):587-597

[7] Ciriminna R, Chavarria-Hernandez N, Hernandez AIR, Pagliaro M. Pectin: A new perspective from the biorefinery standpoint. Biofuels, Bioproducts and Biorefining. 2015;**9**(4):368-377

[8] Peng Q, Xu Q, Yin H, Huang L, Du Y. Characterization of an immunologically active pectin from the fruits of lycium ruthenicum. International Journal of Biological Macromolecules. 2014;**64**:69-75

[9] Babbar N, Dejonghe W, Gatti M, Sforza S, Elst K. Pectic oligosaccharides from agricultural by-products: Production, characterization and health benefits. Critical Reviews in Biotechnology. 2016;**36**(4):594-606

[10] Gomez B, Gullon B, Yanez R, Schols H, Alonso JL. Prebiotic potential of pectins and pectic oligosaccharides derived from lemon peel wastes and sugar beet pulp: A comparative evaluation. Journal of Functional Foods. 2016;**20**:108-121

[11] Li PJ, Xia JL, Nie ZY, Shan Y. Pectic oligosaccharides hydrolyzed from orange peel by fungal multi enzyme complexes and their prebiotic and antibacterial potentials. LWT-Food Science and Technology. 2016;**69**:203-210

[12] Gunter EA, Popeyko OV. Calcium pectinate gel beads obtained from callus cultures pectins as promising systems for colon-targeted drug delivery. Carbohydrate Polymers. 2016;**147**:490-499

[13] Liu L, Fishman ML, Kost J, Hicks KB. Pectin-based systems for colon-specific drug delivery via oral route. Biomaterials. 2003;**24**(19):3333-3343

[14] Sriamornsak P. Application of pectin in oral drug delivery. Expert Opinion on Drug Delivery. 2011;**8**(8):1009-1023

[15] Yapo BM. Pectic substances: From simple pectic polysaccharides to complex pectins—A new hypothetical model. Carbohydrate Polymers. 2011;**86**(2):373-385

[16] Kaya M, Sousa AG, Crepeau MJ, Sorensen SO, Ralet MC. Characterization of citrus pectin samples extracted under different conditions: Influence of acid type and

pH of extraction. Annals of Botany. 2014;**114**(6):1319-1326

[17] Sila DN, Van Buggenhout S, Duvetter T, Fraeye I, De Roeck A, Van Loey A, et al. Pectins in processed fruit and vegetables: Part II–Structure-function relationships. Comprehensive Reviews in Food Science and Food Safety. 2009;**8**(2):86-104

[18] Yoo SH, Lee BH, Lee H, Lee S, Bae IY, Lee HG, et al. Structural characteristics of pumpkin pectin extracted by microwave heating. Journal of Food Science. 2012;**77**(11):C1169-C1173

[19] Gnanasambandam R, Proctor A. Determination of pectin degree of esterification by diffuse reflectance Fourier transform infrared pectroscopy. Food Chemistry. 2000;**68**(3):327-332

[20] Muller-Maatsch J, Bencivenni M, Caligiani A, Tedeschi T, Bruggeman G, Bosch M, et al. Pectin content and composition from different food waste streams. Food Chemistry. 2016b;**201**:37-45

[21] Fraeye I, Duvetter T, Doungla E, Van Loey A, Hendrickx M. Fine-tuning the properties of pectin calcium gels by control of pectin fine structure, gel composition and environmental conditions. Trends in Food Science & Technology. 2010;**21**(5):219-228

[22] Tromp RH, de Kruif CG, Van Eijk M, Rolin C. On the mechanism of stabilisation of acidified milk drinks by pectin. Food Hydrocolloids. 2004;**18**(4):565-572

[23] Voragen AGJ, Coenen GJ, Verhoef RP, Schols HA. Pectin, a versatile polysaccharide present in plant cell walls. Structural Chemistry. 2009;**20**(2):263-275

[24] Bonnin E, Dolo E, Le Goff A, Thibault JF. Characterisation of pectin subunits released by an optimised combination of enzymes. Carbohydrate Research. 2002;**337**(18):1687-1696

[25] Ovodov YS. Current views on pectin substances. Russian Journal of Bioorganic Chemistry. 2009;**35**(3):269-284

[26] Zandleven J, Beldman G, Bosveld M, Schols HA, Voragen AGJ. Enzymatic degradation studies of xylogalacturonans from apple and potato, using xylogalacturonan hydrolase. Carbohydrate Polymers. 2006;**65**(4):495-503

[27] Le Goff A, Renard C, Bonnin E, Thibault JF. Extraction, purification and chemical characterisation of xylogalacturonans from pea hulls. Carbohydrate Polymers. 2001;**45**(4):325-334

[28] Yu LY, Mort AJ. Partial characterization of xylogalacturonans from cell walls of ripe watermelon fruit: Inhibition of endopolygalacturonase activity by xylosylation. Progress in Biotechnology. Vol. 14. 1996. pp. 79-88

[29] Renard C, Crepeau MJ, Thibault JF. Structure of the repeating units in the rhamnogalacturonic backbone of apple, beet and citrus pectins. Carbohydrate Research. 1995;**275**(1):155-165

[30] Maxwell EG, Belshaw NJ, Waldron KW, Morris VJ. Pectin—An emerging new bioactive food polysaccharide. Trends in Food Science and Technology. 2012;**24**(2):64-73

[31] Caffall KH, Mohnen D. The structure, function, and biosynthesis of plant cell wall pectic polysaccharides. Carbohydrate Research. 2009;**344**(14):1879-1900

[32] Yapo BM, Wathelet B, Paquot M. Comparison of alcohol precipitation and membrane filtration effects

on sugar beet pulp pectin chemical features and surface properties. Food Hydrocolloids. 2007;**21**(2):245-255

[33] Coenen GJ, Bakx EJ, Verhoef RP, Schols HA, Voragen AGJ. Identification of the connecting linkage between homo- or xylogalacturonan and rhamnogalacturonan type I. Carbohydrate Polymers. 2007;**70**(2):224-235

[34] Leivas CL, Iacomini M, Cordeiro LM. Pectic type II arabinogalactans from starfruit (Averrhoa carambola L.). Food Chemistry. 2016;**199**:252-257

[35] Palin R, Geitmann A. The role of pectin in plant morphogenesis. Bio Systems. 2012;**109**(3):397-402

[36] Ishii T, Matsunaga T, Pellerin P, O'Neill MA, Darvill A, Albersheim P. The plant cell wall polysaccharide rhamnogalacturonan II self-assembles into a covalently cross-linked dimer. The Journal of Biological Chemistry. 1999;**274**(19):13098-13104

[37] Lara-Espinoza C, Carvajal-Millan E, Balandran-Quintana R, Lopez-Franco Y, Rascon-Chu A. Pectin and pectin-based composite materials: Beyond food texture. Molecules. 2018;**23**(942):1-35

[38] Yapo BM, Robert C, Etienne I, Wathelet B, Paquot M. Effect of extraction conditions on the yield, purity and surface properties of sugar beet pulp pectin extracts. Food Chemistry. 2007;**100**(4):1356-1364

[39] Masmoudi M, Besbes S, Ben Thabet I, Blecker C, Attia H. Pectin extraction from lemon by-product with acidified date juice: Rheological properties and microstructure of pure and mixed pectin gels. Food Science and Technology International. 2010;**16**(2):105-114

[40] Masmoudi M, Besbes S, Chaabouni M, Robert C, Paquot M, Blecker C, et al. Optimization of pectin extraction from lemon by-product with acidified date juice using response surface methodology. Carbohydrate Polymers. 2008;**74**(2):185-192

[41] Shalini R, Gupta DK. Utilization of pomace from apple processing industries: A review. Journal of Food Science and Technology. 2010;**47**(4):365-371

[42] Rascon-Chu A, Martinez-Lopez AL, Carvajal-Millan E, de Leon-Renova NEP, Marquez-Escalante J, Romo-Chacon A. Pectin from low quality 'Golden Delicious' apples: Composition and gelling capability. Food Chemistry. 2009;**116**(1):101-103

[43] Gigli J, Garnier C, Piazza L. Rheological behaviour of low-methoxyl pectin gels over an extended frequency window. Food Hydrocolloids. 2009;**23**(5):1406-1412

[44] Ververis C, Georghiou K, Danielidis D, Hatzinikolaou DG, Santas P, Santas R, et al. Cellulose, hemicelluloses, lignin and ash content of some organic materials and their suitability for use as paper pulp supplements. Bioresource Technology. 2007;**98**(2):296-301

[45] Methacanon P, Krongsin J, Gamonpilas C. Pomelo (citrus maxima) pectin: Effects of extraction parameters and its properties. Food Hydrocolloids. 2014;**35**:383-391

[46] Liu H, Zhu DS, Xu XM, Guo SD, Jin ZY. Rheological, textural and microstructure analysis of the high-methoxy pectin/gelatin mixed systems. Journal of Texture Studies. 2007;**38**(5):577-600

[47] Chen HM, Fu X, Luo ZG. Effect of molecular structure on emulsifying properties of sugar beet pulp pectin. Food Hydrocolloids. 2016;**54**:99-106

[48] Li DQ, Du GM, Jing WW, Li JF, Yan JY, Liu ZY. Combined effects of independent variables on yield and

protein content of pectin extracted from sugar beet pulp by citric acid. Carbohydrate Polymers. 2015;**129**:108-114

[49] Ralet MC, Crepeau MJ, Buchholt HC, Thibault JF. Polyelectrolyte behaviour and calcium binding properties of sugar beet pectins differing in their degrees of methylation and acetylation. Biochemical Engineering Journal. 2003;**16**(2):191-201

[50] Guo XB, Guo XM, Yu SJ, Kong FS. Influences of the different chemical components of sugar beet pectin on the emulsifying performance of conjugates formed between sugar beet pectin and whey protein isolate. Food Hydrocolloids. 2018;**82**:1-10

[51] Funami T, Zhang GY, Hiroe M, Noda S, Nakauma M, Asai I, et al. Effects of the proteinaceous moiety on the emulsifying properties of sugar beet pectin. Food Hydrocolloids. 2007;**21**(8):1319-1329

[52] Colquhoun IJ, Ralet MC, Thibault JF, Faulds CB, Williamson G. Structure identification of feruloylated oligosaccharides from sugar-beet pulp by NMR-spectroscopy. Carbohydrate Research. 1994;**263**(2):243-256

[53] Zaidel DNA, Meyer AS. Biocatalytic cross-linking of pectic polysaccharides for designed food functionality: Structures, mechanisms, and reactions. Biocatalysis and Agricultural Biotechnology. 2012;**1**(3):207-219

[54] Christiaens S, Uwibambe D, Uyttebroek M, Van Droogenbroeck B, Van Loey AM, Hendrickx ME. Pectin characterisation in vegetable waste streams: A starting point for waste valorisation in the food industry. LWT-Food Science and Technology. 2015;**61**(2):275-282

[55] Muller-Maatsch J, Bencivenni M, Caligiani A, Tedeschi T, Bruggeman G, Bosch M, et al. Pectin content and composition from different food waste streams. Food Chemistry. 2016a;**201**:37-45

[56] Dranca F, Oroian M. Extraction, purification and characterization of pectin from alternative sources with potential technological applications. Food Research International. 2018;**113**:327-350

[57] Iglesias MT, Lozano JE. Extraction and characterization of sunflower pectin. Journal of Food Engineering. 2004;**62**(3):215-223

[58] Jimenez A, Rodriguez R, Fernandez-Caro I, Guillen R, Fernandez-Bolanos J, Heredia A. Olive fruit cell wall: Degradation of pectic polysaccharides during ripening. Journal of Agricultural and Food Chemistry. 2001;**49**(1):409-415

[59] Rubio-Senent F, Rodriguez-Gutierrez G, Lama-Munoz A, Garcia A, Fernandez-Bolanos J. Novel pectin present in new olive mill wastewater with similar emulsifying and better biological properties than citrus pectin. Food Hydrocolloids. 2015;**50**:237-246

[60] Stabnikova O, Wang JY & Ivanov V. Value-added biotechnological products from organic wastes. In: Wang LK, Ivanov V, Tay JH, Hung YT editors. Environmental Biotechnology. Totowa-NJ, USA: Humana Press; 2010;**10**:343-394

[61] Kost’alova Z, Hromadkova Z, Ebringerova A. Structural diversity of pectins isolated from the styrian oil-pumpkin (cucurbita pepo var. styriaca) fruit (vol 93, pg 163, 2013). Carbohydrate Polymers. 2014;**99**:831-831

[62] Jun HI, Lee CH, Song GS, Kim YS. Characterization of the pectic polysaccharides from pumpkin peel. LWT-Food Science and Technology. 2006;**39**(5):554-561

[63] Oliveira TIS, Rosa MF, Cavalcante FL, Pereira PHF, Moates GK, Wellner N, et al. Optimization of pectin extraction from banana peels with citric acid by using response surface methodology. Food Chemistry. 2016;**198**:113-118

[64] Banerjee J, Singh R, Vijayaraghavan R, MacFarlane D, Patti AF, Arora A. Bioactives from fruit processing wastes: Green approaches to valuable chemicals. Food Chemistry. 2017;**225**:10-22

[65] Petkowicz CLO, Vriesmann LC, Williams PA. Pectins from food waste: Extraction, characterization and properties of watermelon rind pectin. Food Hydrocolloids. 2017;**65**:57-67

[66] Del Valle M, Camara M, Torija ME. Chemical characterization of tomato pomace. Journal of the Science of Food and Agriculture. 2006;**86**(8):1232-1236

[67] Round AN, Rigby NM, MacDougall AJ, Morris VJ. A new view of pectin structure revealed by acid hydrolysis and atomic force microscopy. Carbohydrate Research. 2010;**345**(4):487-497

[68] Guillon F, Moise A, Quemener B, Bouchet B, Devaux MF, Alvarado C, et al. Remodeling of pectin and hemicelluloses in tomato pericarp during fruit growth. Plant Science. 2017;**257**:48-62

[69] Yapo BM. Lemon juice improves the extractability and quality characteristics of pectin from yellow passion fruit by-product as compared with commercial citric acid extractant. Bioresource Technology. 2009;**100**(12):3147-3151

[70] Pacheco MT, Villamiel M, Moreno R, Moreno FJ. Structural and rheological properties of pectins extracted from industrial sugar beet by-products. Molecules. 2019;**24**:392-408

[71] Jafari F, Khodaiyan F, Kiani H, Hosseini SS. Pectin from carrot pomace: Optimization of extraction and physicochemical properties. Carbohydrate Polymers. 2017;**157**:1315-1322

[72] Korish M. Faba bean hulls as a potential source of pectin. Journal of Food Science and Technology. 2015;**52**(9):6061-6066

[73] Adetunji LR, Adekunle A, Orsat V, Raghavan V. Advances in the pectin production process using novel extraction techniques: A review. Food Hydrocolloids. 2017;**62**:239-250

[74] Puri M, Sharma D, Barrow CJ. Enzyme-assisted extraction of bioactives from plants. Trends in Biotechnology. 2012;**30**(1):37-44

[75] Panouillé M, Thibaul JF, Bonnin E. Cellulase and protease preparations can extract pectins from various plant byproducts. Journal of Agricultural and Food Chemistry. 2006;**54**(23):8926-8935

[76] Yang Y, Wang Z, Hu D, Xiao K, Wu JY. Efficient extraction of pectin from sisal waste by combined enzymatic and ultrasonic process. Food Hydrocolloids. 2018;**79**:189-196

[77] Munarin F, Tanzi MC, Petrini P. Advances in biomedical applications of pectin gels. International Journal of Biological Macromolecules. 2012;**51**:681-689

[78] Sriamornsak P, Wattanakorn N, Takeuchi H. Study on the mucoadhesion mechanism of pectin by atomic force microscopyand mucin-particle method. Carbohydrate Polymers. 2010;**79**:54-59

[79] Ludwig A. The use of mucoadhesive polymers in ocular drug delivery. Advanced Drug Delivery Reviews. 2005;**57**:1595-1639

[80] Glinsky VV, Raz A. Modified citrus pectin anti-metastatic properties: One

bullet, multiple targets. Carbohydrate Research. 2009;**344**:1788-1791

[81] Ferreira-Lazarte A, Kachrimanidou V, Villamiel M, Rastall RA, Moreno FJ. In vitro fermentation properties of pectins and enzymatic-modified pectins obtained from different renewable bioresources. Carbohydrate Polymers. 2018;**199**:482-491

[82] Gullón B, Gullón P, Tavaria F, Pintado M, Gomes AM, Alonso JL, et al. Structural features and assessment of prebiotic activity of refined arabinoxylooligosaccharides from wheat bran. Journal of Functional Foods. 2014;**6**:438-449

[83] Islamova ZI, Ogai DK, Abramenko OI, Lim AL, Abduazimov BB, Malikova MK, et al. Comparative assessment of the prebiotic activity of some pectin polysaccharides. Pharmaceutical Chemistry Journal. 2017;**51**(4):288-291

[84] Munarin F, Giuliano L, Bozzini S, Tanzi MC, Petrini P. Materials Science and Engineering C. 2010;**30**:491-496

[85] Harris R, Nasi W. Gels for Use in Wound Management. World Patent no. WO/2008/015475; 2008

[86] Chan SY, Choo WS. Effect of extraction conditions on the yield and chemical properties of pectin from cocoa husks. Food Chemistry. 2013;**141**(4):3752-3758

[87] Hosseini SS, Khodaiyan F, Yarmand MS. Aqueous extraction of pectin from sour orange peel and its preliminary physicochemical properties. International Journal of Biological Macromolecules. 2016;**82**:920-926

[88] Maran JP, Priya B. Multivariate statistical analysis and optimization of ultrasound-assisted extraction of natural pigments from waste red beet stalks. Journal of Food Science and Technology. 2016;**53**(1):792-799

[89] Andersen NM, Cognet T, Santacoloma PA, Larsen J, Armagan I, Larsen FH, et al. Dynamic modelling of pectin extraction describing yield and functional characteristics. Journal of Food Engineering. 2017;**192**:61-71

[90] Wikiera A, Mika M, Starzynska-Janiszewska A, Stodolak B. Endo-xylanase and endo-cellulase-assisted extraction of pectin from apple pomace. Carbohydrate Polymers. 2016;**142**:199-205

[91] Jeong HS, Kim HY, Ahn SH, Oh SC, Yang I, Choi IG. Optimization of enzymatic hydrolysis conditions for extraction of pectin from rapeseed cake (Brassica napus L.) using commercial enzymes. Food Chemistry. 2014;**157**:332-338

[92] Dominiak M, Sondergaard KM, Wichmann J, Vidal-Melgosa S, Willats WGT, Meyer AS, et al. Application of enzymes for efficient extraction, modification, and development of functional properties of lime pectin. Food Hydrocolloids. 2014;**40**:273-282

[93] Wang M, Huang B, Fan C, Zhao K, Hu H, Xu X, et al. Characterization and functional properties of mango peel pectin extracted by ultrasound assisted citric acid. International Journal of Biological Macromolecules. 2016;**91**:794-803

[94] Koubala BB, Kansci G, Mbome LI, Crepeau MJ, Thibault JF, Ralet MC. Effect of extraction conditions on some physicochemical characteristics of pectins from “Amelioree” and “mango” mango peels. Food Hydrocolloids. 2008;**22**(7):1345-1351

[95] Moorthy IG, Maran JP, Ilakya S, Anitha SL, Sabarima SP, Priya B. Ultrasound assisted extraction of pectin from waste Artocarpus heterophyllus fruit peel. Ultrasonics Sonochemistry. 2017;**34**:525-530

[96] Wang X, Chen QR, Lu X. Pectin extracted from apple pomace and citrus peel by subcritical water. Food Hydrocolloids. 2014;**38**:129-137

[97] Chen HM, Fu X, Luo ZG. Properties and extraction of pectin-enriched materials from sugar beet pulp by ultrasonic-assisted treatment combined with subcritical water. Food Chemistry. 2015;**168**:302-310

[98] Maran J, Prakash K. Process variables influence on microwave assisted extraction of pectin from waste Carcia papaya L. peel. International Journal of Biological Macromolecules. 2015;**73**:202-206

[99] Koubala B, Christiaens S, Kansci G, Van Loey AM, Hendrickx ME. Isolation and structural characterisation of papaya peel pectin. Food Research International. 2014;**55**:215-221

[100] Lefsiha K, Giacomazzab D, Dahmoune F, Mangione MR, Bulone D, San Biagio PL, et al. Pectin from opuntia ficus indica: Optimization of microwave-assistedextraction and preliminary characterization. Food Chemistry. 2017;**221**:91-99

Chapter 4

Role of Pectin in Food Processing and Food Packaging

Thiraviam Vanitha and Mahejibin Khan

Abstract

Pectin is a branched heteropolysaccharide consisting of long-chain galacturonan segments and other neutral sugars such as rhamnose, arabinose, galactose, and xylose. It forms a matrix with celluloses and hemicelluloses and contributes to the cell structure. Due to the presence of several sugar moieties and different levels of methyl esterification, pectin does not have defined molecular weight like other polysaccharides. Pectin has wide applications. It is used as emulsifier, gelling agent, thickener, stabilizer, and fat or sugar replacer in low-calorie foods. Pectin and pectin-derived oligosaccharides can also be used as an important ingredient in functional foods. In recent past, a new application envisaged for pectin polymers as edible films or coating. These films act as natural barrier for exchange of moisture, gases, lipids, and volatiles between food and environment, and protect fruits and vegetable from microbial contamination. The degree of esterification of pectin and other structural modifications defines the functional properties. Herein, various functional properties of pectin in relation to food processing and packaging are discussed.

Keywords: pectin, pectin oligosaccharide, food emulsifier, edible films, functional food, food stabilizer, emulsified food

1. Introduction

Pectin is the major constituent of all plants and makes up approximately two-third of the dry mass of plant primary cell walls. It provides structural integrity, strength, and flexibility to the cell wall and acts as barrier to the external environment [1]. Pectin is also a natural component of all omnivorous diet and is an important source of dietary fiber. Due to the resistant in digestive system and lack of pectin digestive enzymes, human beings are not able to digest pectin directly but microorganism present in large intestine can easily assimilate the pectin and convert it into soluble fibers. These oligosaccharides promote beneficial microbiota in gut and also help in lipid and fat metabolism, glycemic regulation, etc. [2]. Being complex and highly diverse in structure, role of pectin is not only limited to the biological and physiological functions, but it has tremendous potential and contributes substantially in other applications ranging from food processing to pharmaceuticals. Pectin is a water-soluble fiber and used in various food as emulsifier, stabilizer, gelling, and thickening agent.

Commercial pectins are extracted from citrus and apple fruit. On the basis of dry mass, apple pomace contains 10–15% pectin, whereas citrus peel possesses 20–30% pectin. However, pectin has also been extracted in higher amount from several other fruits and their by-products, such as sunflower head, mango peal, soybean hull [3], passion fruit peel [4], sugar beet pulp [5], *Akebia trifoliata* peel [6], peach pomace [7], banana peel [8], chickpea husk [9], and many more [10–23]. **Table 1** summarizes the different types of pectin extracted from various horticultural crops. But detection and extraction of pectin in higher concentration is not sufficient to qualify that fruit as a source of commercial pectin because of the structural variation and modification in side-chain sugars, and also that pectin from different sources has different gelling properties.

S. No	Source	Parts used	Extraction method used	Pectin yield (%)	Type of pectin (HMP/LMP)	Ref
1	Passion fruit	Peel	APP	14.8%	HMP	[4]
2	Banana	Peel	APP	5–21%	HMP (DE, 50–80%)	[8]
3	Chick pea	Husk	Acid extraction, APP, and freeze dried	8%	LMP (DE, 10%)	[9]
4	Krueo Ma Noy	Leaves	APP, DPP	21–28%	LMP (DE, 34–42%)	[11]
5	Yellow Passion	Fruit rind	APP, DPP, MPP	3–16%	HMP (DE, 54–59%)	[12]
6	Durian	Rind	APP	2–10.25%	HMP (DE, 50–64%)	[13]
7	Mulberry	Mulberry bark with epidermis (MBE) and without epidermis (MB)	Extracted using 60–100% isopropanol	11.88%	HMP (MB–DE, 71.13%); LMP (MBE–DE, 24.27%)	[14]
8	Yuzu, citrus family	Pomace	Extracted with APP and enzyme (Viscozyme® L with 1.2×10^{-4} fungal β-glucanase	DPP, APP (7.3–8%)	LMP (APP–DE, 41%; DPP–DE, 46.3%)	[16]
9	Cacao pods	Husk	Extracted with 1 N HNO_3 at different pH and precipitated by ethanol and acetone	3.7–8.6%	LMP (DE 36.7% @ pH 1, DE 44.3% @ pH 3); HMP (DE 52.4% @ pH 2)	[17]
10	Cashew apple	Pomace	AOP at different pH (1.0, 1.5, and 2.0)	10.7–25.3%	LMP (DE, 28–46%)	[18]
11	*Cyclea barbata* Miers (CBM)	Leaves	Extracted with acid and alkali, precipitated the pectin by ethanol	4–8%	HMP (acid treated: 65–75% DE) LMP (Alkali treated: 36% DE)	[19]

S. No	Source	Parts used	Extraction method used	Pectin yield (%)	Type of pectin (HMP/LMP)	Ref
12	Dragon fruit	Peel	Extracted using HCl, precipitated and purified with 70 and 99.6% isopropanol.	18.59%	LMP (DE, 46.95%)	[20]
13	Jackfruit	Peel	Ultrasonic-microwave-assisted extracted (UMAE) pectin	21.5%	HMP (DE, 62.5%)	[22]
14	Potato	Pulp	Extracted with different acids and precipitated by ethanol	4.08–14.34%	LMP (DE, 21.51–37.45%)	[23]

APP, alcohol-precipitated pectin; MPP, metal ion-precipitated pectin; DPP, dialyzed precipitated pectin.

Table 1.
High methoxyl pectins (HMP) and low methoxyl pectins (LMP) from various horticultural crops.

2. Pectin structure

Pectin is a highly complex plant cell wall polysaccharide that plays a significant role in plant growth and development. It is predominantly present in fruits and vegetables and constitutes approximately 35–40% of the primary cell wall in all the dicot plants [24]. The composition and structure of pectin is influenced by the developmental stages of plants [25, 26]. Structural analysis of pectin revealed that it is a polymer comprised of chain-like configuration of approximately 100–1000 saccharide units; therefore, it does not possess a defined structure. In general, pectin is illustrated as a heteropolysaccharide of three components namely, homogalacturonan (HG), rhamnogalacturonan-I (RGI), and rhamnogalacturonan-II (RGII) [28, 29]. The Backbone structure may branch with other neutral sugar chains such as arabinan, xylogalacturonan (XGA), arabinogalactan I (AG-I), and arabinogalactan II (AG-II).

Homogalacturonan (HG) is a polymer of galacturonic acid (GalA), in which Gal A residues are linked together by α-1-4 glycosidic bond and the number of GalA residues in HG may vary from 72 to 100% depending on the source of pectin [30]. For instance, the HG backbone of cashew apple pectin, *C. maxima* pectin, sunflower pectin, citrus pectin, comprises of 69.9–85%, 71–75%, 77–85%, 80–95%, GalA residues, respectively. Amaranth pectin contains more than 80% GalA residues in HG backbone structure. Furthermore, it was also observed that HG may be methoxy-esterified at C-6 and/or O-acetylated at the O-2 and/or O-3. Some exception has also been reported in the detailed structural analysis of HG region of pectin such as C-3 substitution of the galacturonic acids of HG with xylose in pea, apple, carrot, duck weed, etc. [31], and C-2 or C-3 with apiose in duck weeds (*Lemna minor*) [32]. HG is susceptible to both mechanical and enzymatic deesterification and degradation.

Rhamnogalacturonan I represents approximately 20–35% of the pectin polysaccharides. It is the highly branched and heterogeneous polysaccharide which is characterized as repeating units of α-(1 → 2)-linked rhamnose and α-(1 → 4)-linked GalA residues. It can be O-acetylated at O-2 and/or O-3 positions of GalA residues

[33, 34]. Pectin from citrus peels, mung bean, kidney bean, apple fruit, and flax hypocotyls has been reported 100% methyl esterified in the RGI region [35, 36]. The composition of RGI varies in pectin extracted from different sources. In sugar beet pectin, 80 repeating units of [→2] –α-L-Rha-(1–4)- α-D-GalA-(1→) comprised the backbone of rhamnogalacturonan I (RG-I), whereas citrus pectin contains only 15–40 repeating units [37]. The polymeric side chains of galactans and arabinans are substituted at the O-4 position of RG-I backbone. Arabinogalactan I (AG-I) and arabinogalactan II (AG-II) are also reported to be present as polymeric side chains [38–40]. The side chains are often referred to as "hairs" and believed to play an important role in pectin functionality. The loss of side chains may increase the solubility of the pectin [41]. PGI is prone to enzymatic depolymerization. However, protease and acid-catalyzed cleavage of RGI has also been reported [28, 42, 43].

The highly conserved polysaccharide of pectin is rhamnogalacturonan II which constitutes about 10% of the pectin polymer [44]. This polysaccharide is made up of (1 → 4)-linked-α-D-GalA units containing 12 monosaccharide such as apiose, acetic acid, 3-deoxy-manno-2-octulosonic acid (KDO), and 3-deoxy-lyxo-2-heptulosaric acid (DHA) as side chains [30, 39]. GalA present in backbone of rhamnogalacturonan II (RG-II) may be methyl esterified at the C-6 position. The percentage of esterified GalA and acetylated groups in HG chain is termed as the DE and DAc, respectively. It is proposed that in the early developmental stages of plants, highly esterified pectin is formed that undergoes some deesterification in the cell wall or middle lamella. In general, tissue pectin ranges from 60 to 90% DE [45]. Both the DE and the DAc of pectin may vary depending on the method of extraction and plant origin [30, 46]. The functional properties of the pectin are determined by the amount and the distribution of esterified GalA residues in the linear backbone. Presence and distribution of esterified and nonmethylated GalA in pectin define the charge on pectin molecules. Based on their degree of esterification (DE), pectins are classified as high methoxy pectins (HMP) or low methoxy pectins (LMP). DE values of HM pectin range from 60 to 75%, whereas pectin with 20–40% of DE is referred as LM pectin. It was also observed that solubility, viscosity, and gelation properties of pectin are correlated and highly dependent on structural features [47, 48]. Pectin and monovalent salts of pectins are generally soluble in water but di- and trivalent ions are insoluble. The solubility of pectin in water increases with decrease in polymer size and increase in methoxy contents. Pectin powder gets hydrated very fast in water and forms clumps. The solubility of these clumps is very slow. As the pectin molecules come in contact with water, deesterification and depolymerization of pectins start spontaneously. The rate of decomposition of pectin depends on pH and temperature of the solution. As the pH of the solution decreased, with elevated temperature, ionization of carboxylate groups also reduced, which suppresses the hydration and repulsion between the polysaccharide molecules and results in the association of molecules in the form of gels. During thermal processing, solubilization of pectin is affected by β-elimination which depolymerized the pectin molecule and reduced its chain length. Small polymers have poor affinity with cell wall framework and solubilize easily. However, preheating, as well as reduced moisture contents in thermal processing, adversely affects the solubility of pectin in water [49, 50].

3. Pectin as food emulsifier

Food additives that are used in food processing to blend two immiscible liquids to produce a desirable product are known as food emulsifier or emulgent. These additives act as surface-active agents on the border of immiscible layers and reduce oil crystallization and prevent water separation. Emulsifiers are used in large number of food products such as ice creams, low-fat spreads, yoghurts, margarine, salad dressings, salty spreads, bakery products, and many other creamy sauces, to keep them in stable emulsion [27]. Emulsifiers increase the whip-ability of batters, enhance mouthfeel of the products, and improve texture and shape of the dough. Moreover, emulsions also help to encapsulate the bioactives [51]. Based on the disperse phase, there are two types of emulsion: oil in water (O/W) and water in oil (W/O). Milk, mayonnaise, dressings, and various beverages are some examples of O/W emulsion, whereas butter and margarine are the typical examples of W/O emulsion. Progress in hydrocolloid chemistry has resulted in the development of multitype emulsion such as O/W/O and O/W/O type emulsion (**Figure 1**). These emulsions are very important for fat reduction or encapsulation of bioactives and are used in preparation and stabilization of various low-fat creams, seasoning, and flavoring of sauces [52].

Commonly used emulsifiers in food processing are (i) small-molecular surfactant such as lectithins, derivatives of mono- and diglycerides prepared by mixing edible oils with glycerin or ethylene oxide, fatty acid derivatives such as glycol esters, sorbitan esters, polysorbates and (ii) macromolecular emulsifiers that include proteins and plant-based polymers such as soy polysaccharide, guar gum, modified starch, pectin, etc. [53]. As far as the properties of food emulsifier are concern, a good emulsifier should be low in molecular weight, capable to reduce the surface tension rapidly at interface, and should be soluble in continuous phase [54]. Research on food additives revealed the adverse effect of synthetic food additives on human being. Chassaing et al. found that polysorbate 80(P80) or carboxy methyl cellulose (CMC) had adverse effects on gut microbiota and their continuous use triggered the weight gain and metabolic syndrome after 12 weeks of administration in mouse [55]. A recent research carried out on mice shows that regular use of P80 and CMC triggers low-grade intestinal inflammation which may ultimately lead to the development of colon cancer [56]. Therefore, safety issues with the synthetic

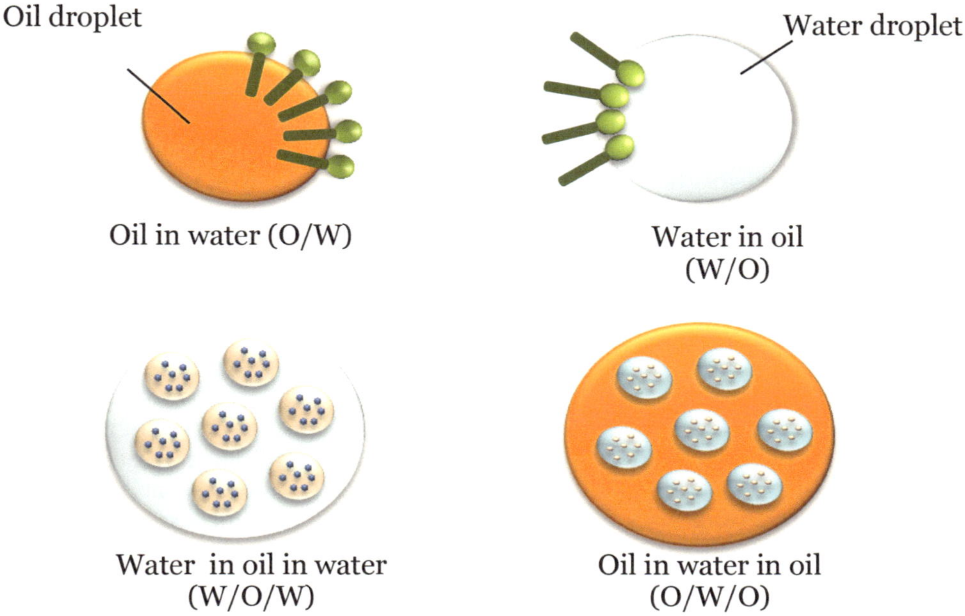

Figure 1.
Types of emulsions.

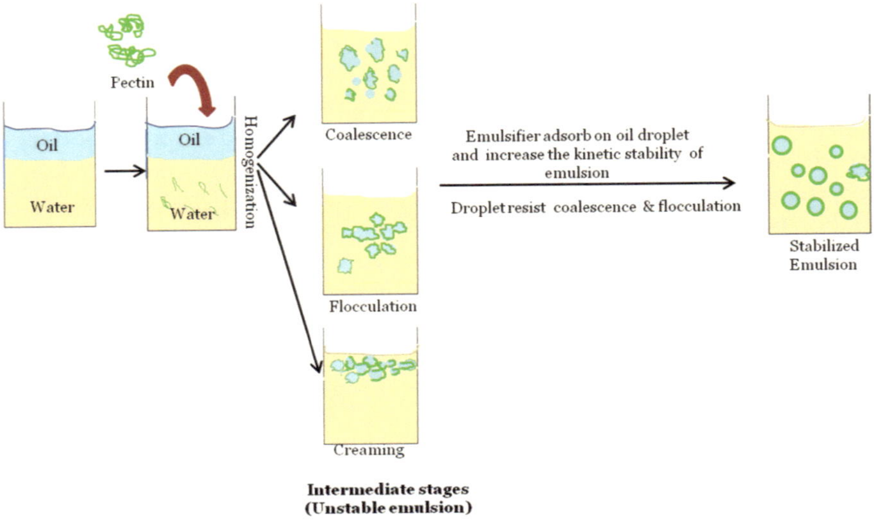

Figure 2.
Emulsion formation and stabilization using polymer as emulgent.

food additives and consumer's demand for all natural food ingredients have necessitated the use of plant-based emulsifiers and stabilizers in food.

Pectin is a natural hydrocolloid which exhibits wide spectrum of functional properties. Because of the gelling ability of pectin, it is used as viscosity enhancer. During emulsification process, pectin molecules adsorb at the fine oil droplets from at O/W interface and protect the droplet from coalescing with adjacent drops (short-term stability). The quality of emulsifier is defined by its ability to provide long-term stability against flocculation and coalescence [27]. **Figure 2** depicts the stages in long-term emulsion formation using pectin as emulgent. When the viscosity of the continuous phase is increased, the movements of oil droplets become restricted which improves the shelf life of emulsion [57]. In the past decade, some pectin has also been reported to exhibit surface active behavior in oil-water interface and thereby stabilizing the fine oil droplets in emulsion [42, 58]. These functions of pectin are determined by its source, structural modification during processing, distribution of functional groups in pectin backbone, and also by various extrinsic factors such as pH, temperature, ionic strength, cosolute concentration, etc. The emulsification or surface active properties of pectin, i.e., formation of fine oil droplets, are mainly contributed due to the high hydrophobicity of protein residue present in pectin [46, 59] and also by hydrophobic nature of acetyl, methyl, and feruloyl esters [42, 60], whereas emulsion-stabilizing ability is attributed to the carbohydrate moieties and their conformational features [61].

3.1 Mechanism of emulsion formation and stabilization

The mechanism of emulsion formation is shown in **Figure 3**. Different models explain the emulsion formation as covalently bound protein moieties in pectin are adsorbed onto the oil-water interface [46], form anchor points at the interface, and reduce the interfacial tension while the charged carbohydrate units extend into the aqueous phase [62] and stabilize by steric and viscosity effects in the aqueous phase(**Figure 3a**). Now, it is a well-established fact that pectin from different source shows variability in structure and protein contents. Leroux et al. identified many anchor points in sugar beet pectin (SBP) molecules [46], and proposed a loop-and-tail model (**Figure 3b**). According to the authors, only a limited amount of protein

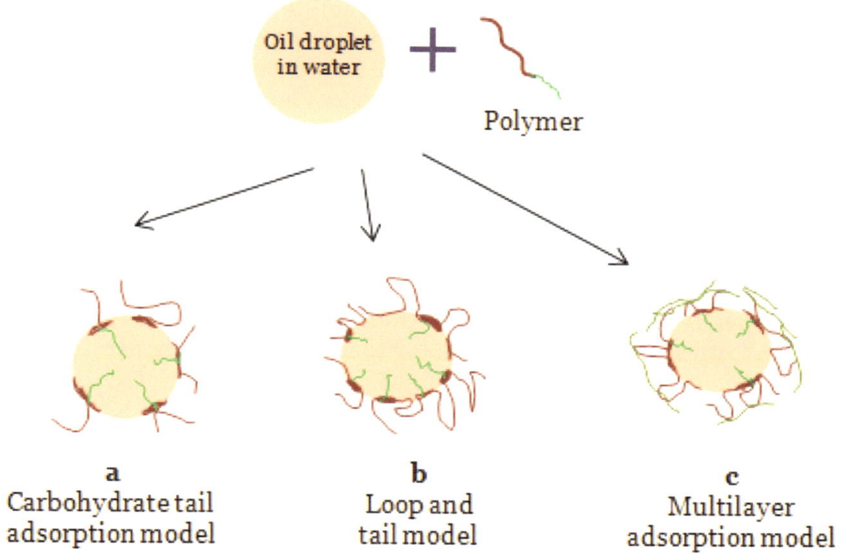

Figure 3.
Different models showing pectin adsorption at oil/water interface during emulsion formation.

is adsorbed at the oil surface and acts as main moiety in the stabilization of the emulsion. This model was further confirmed by Siew and others [62]. The study was carried out to measure the thickness of the adsorbed SBP on oil-water interface layer, proposed a multilayer adsorption model (**Figure 3c**). Electrostatic interactions between the positively charged protein moiety and the negatively charged carbohydrate moiety were also reported.

Pectin O/W emulsion is generally stabilized through steric and electrostatic interaction. The carbohydrate moieties and neutral sugar side chains of RG I region of pectin confer the stability to the pectin emulsions through steric properties of the adsorbed polymers, when pectin is used as monoemulsifiers. In addition, pectin reversible association with galactan/arabinogalactan prior to emulsification also improves the emulsion stability [42, 63]. Electrostatic stabilization of emulsion is ascribed to sugar moieties and structural features of the HG units of pectin. If the pH of dispersion medium is above 3.5, nonmethylated carboxylic group of HG region gets ionized and confers charge on the pectin surface. Interaction of an ionic surfactant with oil droplets results in electrostatic stabilization [64]. Pectin viscosity also plays an important role in controlling the emulsion stability. HG region-rich pectin shows higher intrinsic viscosity ($[\eta]$); therefore, HG and RG ratio of pectin and molecular interactions that improve the intrinsic viscosity ($[\eta]$) of pectin solution also contributes in shelf life of emulsion [65, 66]. It has also observed that structural features of pectin such as pectin protein content, molecular mass, and presence of ferulic acid, and acetyl group in carbohydrate moieties of pectin also affect pectin's emulsifying and emulsion stabilization properties [15]. Williams et al. showed that ferulic acid-rich pectin did not show significant difference in emulsifying ability of pectin when compared with pectin poor in ferulic acid [67]. Digestion of sugar beet pectin(SBP) with acidic proteases resulted in formation of larger size of oil droplet, lower creaming stability, and loss of emulsifying activity of SBP which confirms that protein contents of SBP play an important role in emulsifying ability of the polymer [42]. Nevertheless, in other research, it was also found that protein-rich fractions of SBP did not necessarily displayed better emulsifying ability; therefore, it was concluded that both protein with carbohydrate moiety together help in controlling emulsifying ability of SBP. Castellani et al. further suggest that both the carbohydrate and protein moieties function together as unit and affect the hydrophilic-hydrophobic equilibrium of the SBP molecule [68]. Therefore, when SBP is digested with proteases or other enzyme, a single moiety may function differently. Furthermore, it was also proposed that protein folding may also mask the hydrophobic effect of protein and thus affect the overall properties of the polymers [69].

Molecular weight of pectin has also been reported to affect the emulsifying capacity of pectin. Pectin with low molecular weight was more efficient in stabilizing small emulsion droplets than high-molecular weight pectin. However, very small size of citrus pectin had negative effect on emulsion-stabilizing ability of pectin. It could be due to the poor steric stabilization of depolymerized polymer [59].

3.2 Pectin-containing emulsion-based food

Emulsion-based food products can be defined as a network of pectin-protein molecules entrapping the oil droplet in between. Nowadays, a large number of pectin- and polysaccharide-based emulsified low-fat dairy products, meat products, spreads or desserts, bakery products, sauces, etc., are available in market. Low-fat and low-cholesterol mayonnaise, low-fat cottage cheese, low-fat drinking yogurt, and flavored oil-containing acidified milk drinks are the few examples of pectin-based emulsified products. These products are prepared by replacing full-fat milk from skimmed milk, emulsified oil, and whey proteins [70, 71]. A low-fat cheese was prepared using skimmed milk and water-in-oil-in-water ($W_1/O/W_2$) emulsified canola oil. Different emulsifiers such as amidated low-methoxyl pectins (LMP), gum arabic (GA), carboxymethylcellulose (CMC), and combinations of GA-CMC or GA-LMP were used to stabilize the emulsion. Textural characteristics and sensory evaluation of low-fat cheese show that polymers used to stabilize the emulsion affected both microcrystalline structure and organoleptic properties. The cheese prepared using GA and LMP was almost similar in textural characteristics to the full-fat milk cheese [72]. In another study, Liu et al. compared the textural and structural features and sensory quality of full-fat and low-fat cheese analogs prepared with or without the incorporation of pectin [71]. Microstructure analysis using scanning electron microscopy revealed that full-fat cheese was denser and contained higher concentration of fat globules than low-fat cheese made with or without pectin. Comparison within the low-fat cheese analogs showed clear difference in their hardness, gumminess, chewiness, and adhesiveness. Addition of pectin had positive effect on textural and sensory attribute and scored better in mouthfeel also.

Low-fat (Lf) mayonnaise was prepared by partial replacement of egg yolk and incorporation of pectin as emulsifier [73, 74]. Pectin weak gel, pectin micro-encapsulation, and whey protein isolate were used in preparation of low-fat (Lf) mayonnaise. Physicochemical and sensory properties of Lf mayonnaise were compared with full-fat (Ff) mayonnaise; Lf mayonnaise had low energy and more water contents than Ff. Textural features and rheological properties of the Lf and Ff mayonnaise were similar and both displayed thixotropic shear thinning behavior and categorized as weak gels. Moreover, Lf mayonnaise prepared using pectin had better acceptability than whey protein incorporation [75]. Emulsified oil is used as an effective delivery system of active compound in functional foods, and also serves as milk fat replacer in fat-free dairy products. To improve the nutritional value of food, low-fat dairy products are produced, whereas saturated milk fat is generally replaced with emulsified-unsaturated vegetable oils [76].

In recent year, pectin in combination with inulin has been reported to prepare low-fat meat batter. Méndez-Zamora et al. studied the effect of substitution of animal fat with different formulations of pectin and inulin on chemical composition, textural, and sensory properties of frankfurter sausages [77]. Finding of the research showed that fracturability, gumminess, and chewiness of the low-fat sauces were slightly lower than those of the control. However, addition of 15% inulin improves the sensory properties. In a similar work, replacement of pork back fat with 15% pectin and 15% inulin was found effective in maintaining the physico-chemical properties and emulsion stability of the low-fat meat batter [78].

4. Pectin as gelling agent

The use of pectin in food products as a gelling agent is a long tradition. Later on, it was discovered that pectin forms different types of viscoelastic solution under suitable conditions. This property of pectin is commercially exploited in preparation of jams, jellies, and marmalades. Rheological behaviors of pectin depend on pectin source, its degree of methylation, distribution of nonmethylated GalA unit on pectin backbone, and degree of acetylation, and also on various extrinsic factors such as temperature, pH, concentration, and presence of divalent ions. At a constant pH, the setting time of pectin increases with decreasing DM and degree of blockiness (DB) in the absence of bivalent ions [79]. Therefore, on the basis of gelling process, pectin is classified as rapid, medium, and slow set pectin [80].

Gelling process of pectin and its stabilization follows different mechanisms for different types of pectin. HMP form gels in a narrow pH range (2.0–3.5) in the presence of sucrose at a concentration higher than 55% w/v in medium. During the gelatin process of HMP, junction zones are formed due to the cross-linking of two or more pectin molecules. These junctions are stabilized by weak molecular interaction such as hydrogen and hydrophobic bonds between polar and nonpolar methyl-esterified groups and require high sugar concentration and low pH [81]. These gels are thermally reversible. LMP can form gel over a wide pH range (2.0–6.0) independent of sucrose, but requires divalent ion, such as calcium [82, 83]. LMP follow the eggbox model for its gelation, where positively charged calcium ions (Ca^{2+}) are entrapped in between the negatively charged carboxylic group of pectin. The zigzag network of Ca^{2+} ion and GalA molecules looks like eggbox, and therefore, model is named as eggbox model [80]. These gels are stabilized by electrostatic bonds. In the presence of Ca^{2+}, calcium bridges are formed with pectin molecules that make the solution more viscous. At the higher pH, the ionic strength of the solution is increased and thus more Ca^{2+} is needed for gelation. In case of highly acetylated pectin such as sugar beet, acetyl groups cause steric hindrances and interfere with the Ca^{2+} ion and GalA bond formation, thus preventing gel formation. Kuuva et al. [84] reported that enzymatic modification in pectin structure, i.e., removal of acetyl groups using α-arabinofuranosidase (α-Afases) and acetyl esterase enzymes, can improve the gelling property of acetylated pectin.

HMP are generally used in preparation of standard jams where sugar contents are above 55%, high-quality, tender confectionary jellies, fruit pastes, etc. LMP do not require sugar for its gelatin and therefore preferred choice for the production of low-calorie food products such as milk desserts, jams, jellies, and preserves, [28, 85]. LM pectins are more stable in low pH and high temperature conditions as compare to HM pectins and can be stored for more than a year.

5. Pectin in food packaging

Food packaging is one of the fastest growing segments of food industry. Traditionally, packaging system was limited to the containers and packaging material to transport the food items from manufacturer to the retail market and then to the consumers. Such type of packaging was unable to contribute in the extension of the shelf life and maintenance of the quality of the products. Due to the globalization of food market and increasing demand of shelf-stable processed food that retains the natural properties of food, the need of functional/active packaging material is increasing. To meet the industrial demand, a number of polymers are being synthesized and used in food packaging because of their flexibility, versatility, and cost effectiveness. Although, synthetic materials are able to fulfill all the industrial needs and keep food fresh and safe by protecting them from abiotic factors such as moisture, heat, oxygen, unpleasant odor, and biotic components such as micro- and macroorganisms. But, disposal of nonbiodegradable packaging material is a serious problem which poses a threat to the environment. Therefore, more research has been focused on the development of biodegradable packaging for food packaging applications using poly(lactic acid) (PLA), poly(hydroxyalkanoates) (PHAs), starch, etc. [86]. Among all the natural polymers, polysaccharides are gaining more attention as they are versatile in nature and easily available in relatively low cost.

A variety of natural polysaccharides, such as pectin, chitosan derivatives, alginate, cellulose, seaweed extract, and starch are usually used in the preparation of edible films and coatings [87]. Pectin is one of the most significant renewable natural polymers which are the main component of all the biomass and ubiquitous in nature. Being flexible in nature, pectin and its derivatives are used in many biodegradable packaging materials that serve as moisture, oil, and aroma barrier, reduce respiration rate and oxidation of food [88]. Pectin along with food grade emulsifiers is also used in the preparation of edible films. These films are used in fresh and minimally processed, fruits and vegetables, foods and food products as pectin is the main component of the omnivorous diet and can be metabolized. Edible coating protects the nutritional properties of the food and also saves highly perishable food from the enzymatic browning, off-flavor development, aroma loss, retards lipid migration, and reduces pathogen attack during storage.

At low pH, LM pectins are cross-linked with calcium cations and form hard gels. These gels have highly stable structure and act as water barriers. Because of these properties, LM pectin films are used as edible coatings [88, 89]. Extension of shelf life of avocado fruits was also reported to over a month at 10°C by using edible pectin films. It was found that when avocados were coated with edible pectin films and stored at 10°C, rate of oxygen absorption and rate of respiration decreased which results in delaying of texture and color change of fruits [90]. Oms-Oliu et al. used calcium chloride and sunflower oil cross-linked with LM pectin films onto fresh-cut melon to see the effect on extension of shelf life of cut fruits [91]. It was observed that edible pectin films maintained the initial firmness, decrease the wounding stress of fresh-cut fruits, and prevent the dehydration during storage up to 15 days at 4°C but could not reduce the microbial growth onto the fresh melon. It has been observed that to reduce the respiration rate and to prevent the off-flavor development, different pectin and emulsifier formations are required for different fruits. Edible coating film formulation consisted on pectin, sorbitol, and bee wax was successfully used by Moalemiyan et al. to keep the fresh-cut mangoes in original state for over 2 weeks [92]. Whereas in a similar study, pectin coating containing sucrose and calcium lactate was able to prevent the fruits' respiration rate and maintain sensory properties in fresh melon fruits for up to 14 days storage at 5°C. In a similar study [93], pectin edible coating solution containing pectin (3%), glycerol (2.5%), polyvinyl alcohol (1.25%), and citric acid (1%) was prepared and applied on sapota fruits by dipping method and uncoated sapota fruits were used as control. Both the treated and control fruits were stored at 30 ± 3°C. Physicochemical parameters namely, weight, color, firmness, acidity, TSS, pH, and ascorbic acid contents of both the coated and control fruits were measured at regular interval up to 11th day of the storage at 30 ± 3°C. Reduced rate of change in weight loss and other parameters were reported in pectin-coated sapota as compared to control fruits and it was observed that pectin film formulation was able to maintain good quality attributes and extend the shelf life of pectin-coated sapota fruits up to 11 days of storage at room temperature, whereas control fruits were edible up to 6 days. Furthermore, it was also observed that sapota fruits dipped in sodium alginate containing 2% pectin solution for 2 min were more effective in maintaining the organoleptic properties up to 30 days of refrigerated storage as compared to sapota fruits dipped for 4 min and untreated sapota fruits [94]. Bayarri et al. developed antimicrobial films using lysozyme and LM pectin complex. The main purpose of the study was to control the release of lysozyme in packaged food and to target lysozyme-sensitive bacteria such as *Bacillus* and *Clostridium.* It was observed that in the presence of fungal pectinase, due to the dissociation of pectin linkage, lysozyme activity of films increased remarkably. Many food-contaminating bacteria are pectinase producing and such type of films may be used to control food contaminants. These results have opened new avenues for custom-made biodegradable film [95].

In last few years, some researchers have focused on pectin-based coating containing edible essential to improve the antimicrobial properties and to enhance the efficiency of the pectin films. Edible coating formulation containing sodium alginate and pectin (PE) enriched with eugenol (Eug) and citral (Cit) essential oil at different concentrations was used to increase the shelf life of strawberries. Physical and organoleptic parameters of coated fruits stored at 10°C for 14 days show that formulation containing PE 2% + Eug 0.1%; PE 2% + Cit 0.15% was more suitable than sodium alginate-based formulations [96]. Pectin coating containing lemon and orange peel essential oils was reported to increase the shelf life and quality attributes of the strawberry fruits up to 12 days when stored at 5°C. It was also observed that fruits coated with pectin + 1% orange essence showed less weight loss and soluble solids as compare to their control during the storage [97]. Sanchís et al. studied the combined effect of edible pectin coating with active modified atmospheric packaging on fresh-cut "Rojo Brillante" persimmon. Persimmon fruit slices were coated by dipping in the pectin-based emulsion or in water as control. Both the treated and control slices were packed under 5 kPa O_2 (MAP) or under ambient atmosphere for up to 9 days at 5°C. Various parameters, such as package gas composition, color and firmness of slice, polyphenol oxidase activity, were measured during storage. It was observed that edible coating along with MAP significantly reduced the CO_2 emission and O_2 consumption in the packaged fruits. Furthermore, coating was also effective in controlling microbial growth and reducing enzymatic browning and maintains good sensory parameters up to 10 days on storage [98].

Drying is the traditional and oldest method of fruit and vegetable preservation. It decreases the enzymatic activity, reduces the moisture contents, and protects the food from microbial attack. However, drying results in loss of nutrients, vitamins, heat-labile enzymes, modifies the texture, color, and organoleptic quality of dried fruits and vegetables and therefore diminishes the market value also. Pretreatment of food products with pectin coatings containing other bioactive compound such as ascorbic acid, $CaCl_2$, edible gum, etc., before drying or blanching has been proposed as an effective method to preserve the nutritional as well as organoleptic quality of dried food [99]. Recent researches have shown that application of pectin coating could protect the moisture and vitamin C loss in pretreated papaya slice and osmotic dehydrated pineapple. In one of the research [100], pineapple slice was pretreated with pectin coating formulation containing (50%)/calcium lactate (4%)/ascorbic acid (2%) solutions and then dried by hot-air-drying method. Physicochemical analysis of dried product showed less reduction in vitamin C contents as compared to untreated pineapple slice. In a similar work, pectin coating supplement with vitamin C (1%) was used for precoating of papaya slice. It was found that incorporation of vitamin C did not affect the drying process. However, significant increase in vitamin C content was observed in final product [101].

Frying is a method of cooking that causes changes in chemical and physical parameters of food and enhances the taste. However, high temperature vaporizes the water of food and affects the nutritional properties due to protein denaturation and starch gelatinization. The oil uptake during frying is affected by various parameters such as type of oil used, frying temperature and duration, product moisture content, shape, porosity, prefrying treatment, etc. [102]. Surface area and pretreatment of products are the major factors that determine the oil absorbed. Edible coating has also been used successfully, to reduce the oil uptake during frying in various deep-fried products. Reduction in oil uptake and improvement of texture and quality of potato slices was reported by Daraei Garmakhany et al. in 2008. Authors found that coating of potato slices with pectin, guar, and CMC solutions can reduce the oil uptake when compared with nontreated potato chips [103]. Similar results were also obtained by Khalil, where a combination of pectin or sodium alginate with calcium chlorides significantly reduces the oil uptake of French fries. Coating formulation of 0.5% calcium chloride and 5% pectin was most effective in reducing the oil uptake [104]. Kizito et al. used different edible coatings (pectin, carboxy methyl cellulose, agar, and chitosan) at a concentration of 1–2% for pretreatment of potato chips, followed by deep frying of chips. Fried chips were analyzed biochemically and organoleptically to investigate the quality attributes of the products. It was revealed that all the coating polymers were successful in reducing the oil uptake but pectin was most effective and reduced oil uptake up to 12.93%, followed by CMC (11.71%), chitosan (8.28%), and agar (5.25%) and significantly improved moisture retention of strips ($p < 0.05$) [105].

6. Conclusion

The application of natural polymers in food industry is increasing day by day. Researchers are focusing more and more toward the pectin because of the ease-of-availability, structural flexibility, and versatile composition. Pectin can be sourced from a number of easily available horticulture crops (**Table 1**). Pectin is a hydrocolloid which is used as a food emulsifier, gelling agent, thickener, and stabilizer. It is the preferred choice of most of the food processors as fat or sugar replacer in low-calorie foods. In the recent years, increasing demand of ready-to-serve foods, fresh-cut fruits, and vegetable has opened a new market for edible films. Being biodegradable and recyclable, a lot of research is being done on pectin-based edible film formulations. These films reduce the exchange of moisture, gases, lipids, and volatiles between food and environment, and also serve as protective barrier for microorganisms.

Even though a lot of information is available regarding pectin structure and many pectin-based products are available in market, role of many carbohydrate moieties and their effect on various function of pectin are not yet well defined. Therefore, it is necessary to understand the structural-function relationship of pectin and its interactions for developing functional food products.

Acknowledgements

The authors thank Director, CSIR-CFTRI for the encouragement.

DOI: http://dx.doi.org/10.5772/intechopen.83677

Conflict of interest

The authors declare no conflict of interest.

Author details

Thiraviam Vanitha[1] and Mahejibin Khan[2*]

1 Department of Fruits and Vegetable Technology, CSIR-Central Food Technological Research Institute, Mysore, India

2 CSIR-Central Food Technological Research Institute-Resource Centre, Lucknow, India

*Address all correspondence to: mahejibin@cftri.res.in

References

[1] Talbott LD, Ray PM. Molecular size and separability features of pea cell wall polysaccharides. Plant Physiology. 1992;**92**:357-368. DOI: 10.1104/pp.98.1.357

[2] Khan M, Ekambaram N, Umesh-Kumar S. Potential application of pectinase in developing functional foods. Annual Review of Food Science and Technology. 2013;**4**:21-34. DOI: 10.1146/annurev-food-030212-182525

[3] Monsoor MA, Proctor A. Preparation and functional properties of soy hull pectin. Journal of the American Oil Chemists' Society. 2001;**78**(7):709-713. DOI: 10.1007/s11746-001-0330-z

[4] Kulkarni SG, Vijayanand P. Effect of extraction conditions on the quality characteristics of pectin from passion fruit peel *(Passiflora edulis f. flavicarpa* L.). LWT- Food Science and Technology. 2010;**43**:1026-1031. DOI: 10.1016/j.lwt.2009.11.006

[5] Levigne S, Ralet MC, Thibault JF. Characterization of pectins extracted from fresh sugar beet under different conditions using an experimental design. Carbohydrate Polymers. 2002;**49**:145-153. DOI: 10.1016/S0144-8617(01)00314-9

[6] Jiang YDY, Zhu X, Xiong H, Woo MW, Hu J. Physicochemical and comparative properties of pectins extracted from Akebia trifoliata var. australis peel. Carbohydrate Polymers. 2012;**87**:1663-1669. DOI: 10.1016/j.carbpol.2011.09.064

[7] Faravash RS, Ashtiani FZ. The influence of acid volume, ethanolto-extracts ratio and acid-washing time on the yield of pectic substances extraction from peach pomace. Food Hydrocolloids. 2008;**22**:196-202. DOI: 10.1016/j.foodhyd.2007.04.003

[8] HappiEmaga T, Ronkart SN, Robert C, Wathelet B, Paquot M. Characterisation of pectins extracted from banana peels (Musa AAA) under different conditions using an experimental design. Food Chemistry. 2008;**108**:463-471. DOI: 10.1016/j.foodchem.2007.10.078

[9] Urias-Orona V, Rascon-Chu A, Lizardi-Mendoza J, Carvajal-Millan E, Gardea A, Ramírez-Wong B. A novel pectin material: Extraction, characterization and gelling properties. International Journal of Molecular Sciences. 2010;**11**:3686-3695. DOI: 10.3390/ijms11103686

[10] Pagán J, Ibarz A, Llorca M, Pagán A, Barbosa-Cánovas GV. Extraction and characterization of pectin from stored peach pomace. Foodservice Research International. 2001;**34**:605-612. DOI: 10.1016/S0963-9969(01)00078-3

[11] Singthong J, Cui SW, Ningsanond S, Douglas Goff H. Structural characterization, degree of esterification and some gelling properties of Krueo Ma Noy (*Cissampelos pareira*) pectin. Carbohydrate Polymers. 2004;**58**:391-400. DOI: 10.1016/j.carbpol.2004.07.018

[12] Yapo BM. Pectin quantity, composition and physicochemical behaviour as influenced by the purification process. Food Research International. 2009;**42**:1197-1202. DOI: 10.1016/j.foodres. 2009.06.002

[13] Wai WW, Alkarkhi AFM, Azh Easa AM. Effect of extraction conditions on yield and degree of esterification of durian rind pectin: An experimental design. Food and Bioproducts Processing. 2010;**88**:209-214. DOI: 10.1016/j.fbp.2010.01.010

[14] Liu H, Xu XM, Guo SD. Comparison of full-fat and low-fat cheese analogues

with or without pectin gel through microstructure, texture, rheology, thermal and sensory analysis. International Journal of Food Science and Technology. 2008;**43**:1581-1592. DOI: 10.1111/j.1365-2621.2007.01616.x

[15] Karnik D, Wicker L. Emulsion stability of sugar beet pectin fractions obtained by isopropanol fractionation. Food Hydrocolloids. 2017;**74**. DOI: 10.1016/j.foodhyd.2017.07.041

[16] Lim J, Yoo J, Ko S, Lee S. Extraction and characterization of pectin from Yuza (*Citrus junos*) pomace: A comparison of conventional-chemical and combined physical-enzymatic extractions. Food Hydrocolloids. 2012;**29**:160-165. DOI: 10.1016/j.foodhyd.2012.02.018

[17] Yapo BM, Koffi KL. Extraction and characterization of gelling and emulsifying pectin fractions from cacao pod husk. Journal of Food and Nutrition Research. 2013;**1**(4):46-51. DOI: 10.12691/jfnr-1-4-3

[18] Yapo BM, Koffi KL. Extraction and characterization of highly gelling low methoxy pectin from cashew apple pomace. Food. 2014;**3**(1):1-12. DOI: 10.3390/foods3010001

[19] Mackaman P, Tangsuphoom N, Chavasit V. Effect of extraction condition on the chemical and emulsifying properties of pectin from Cyclea barbata Miers leaves. International Food Research Journal. 2014;**21**(2):799-806

[20] Rahmati S, Abdullah A, Momeny E, Kang OL. Optimization studies on microwave assisted extraction of dragon fruit (*Hylocereus polyrhizus*) peel pectin using response surface methodology. International Food Research Journal. 2015;**22**(1):233-239

[21] Maskey B, Dhakal D, Pradhananga M, Shrestha NK. Extraction and process optimization of bael fruit pectin. Food Science & Nutrition. 2018;**6**:1927-1932. DOI: 10.1002/fsn3.761

[22] Xu SY, Liu JP, Huang X, Du LP, Shia FL, Dong R, et al. Ultrasonic-microwave assisted extraction, characterization and biological activity of pectin from jackfruit peel. LWT- Food Science and Technology. 2018;**90**:577-582. DOI: 10.1016/j.lwt.2018. 01.007

[23] Yang JS, Mu TH, Ma MM. Extraction, structure, and emulsifying properties of pectin from potato pulp. Food Chemistry. 2018;**244**:197-205. DOI: 10.1016/j.foodchem.2017.10.059

[24] Van Buggenhout S, Sila DN, Duvetter t VLA, Autio K, Hendrickx M. Pectins in processed fruits and vegetables: Part III—Texture engineering. Comprehensive Reviews in Food Science and Food Safety. 2009;**8**:105-117. DOI: 10.1111/j.1541-4337.2009.00072.x

[25] Huyskens-Keil S, Prono-Widayat H, Lüdders P, Schreiner M. Post-harvest quality of Pepino (*Solanum muricatum* ait.) fruit in controlled atmosphere storage. Journal of Food Engineering. 2006;**77**:628-634. DOI: 10.1016/j.jfoodeng.2005.07.028

[26] Van Linden V, Sila DN, Duvetter T, De Baerdemaeker J, Hendrickx M. Effect of mechanical impact-bruising on polygalacturonase and pectinmethylesterase activity and pectic cell wall components in tomato fruit. Postharvest Biology and Technology. 2008;**47**:98-106. DOI: 10.1016/j.postharvbio.2007.06.006

[27] ED Ngouémazong, Christiaens S, Shpigelman A, Van Loey AM, Hendrickx ME. The emulsifying and emulsion-stabilizing properties of pectin: A review. Comprehensive Reviews in Food Science and Food Safety. 2015;**14**:705e718. DOI: 10.1111/1541-4337.12160

[28] Thibault J, Renard C, Axelos M, Roger P, Crepeau M. Studies of the length of homogalacturonic regions in pectins by acid hydrolysis. Carbohydrate Research. 1993;**238**:271-286. DOI: 10.1016/0008-6215(93)87019-O

[29] Zhan D, Janssen P, Mort A. Scarcity or complete lack of single rhamnose residues interspersed within the homogalacturonan regions of citrus pectin. Carbohydrate Research. 1998;**308**:373-380. DOI: 10.1016/S0008-6215(98)00096-2

[30] Voragen AGJ, Coenen GJ, Verhoef RP, Schols HA. Pectin, a versatile polysaccharide present in plant cell walls. Structural Chemistry. 2009;**20**:263-275. DOI: 10.1007/s11224-009-9442-z

[31] Renard CM, Weightman RM, Thibault J. Xylose-rich pectins from pea hulls. International Journal of Biological Macromolecules. 1997;**21**:155-162. DOI: 10.1016/S0141-8130(97)00055-X

[32] Longland J, Fry SC, Trewavas A. Developmental control of apiogalacturonan biosynthesis and UDP-apiose production in duck weed. Plant Physiology. 1989;**90**:972-976. DOI: 10.1104/pp.90.3.972

[33] Komalavilas P, Mort AJ. The acetylation of O-3 of galacturonic acid in the rhamnose-rich portion of pectins. Carbohydrate Research. 1989;**189**:261-272. DOI: 10.1016/0008-6215(89)84102-3

[34] Vincken J, Schols H, Oomen R, McCann M, Ulvskov P, Voragen A, et al. If homogalacturonan were a side chain of rhamnogalacturonan I. implications of cell wall architecture. Plant Physiology. 2003;**132**:1781-1789. DOI: 10.1104/pp.103.022350

[35] Ridley BL, O'Neil MA, Mohnen D. Pectins: Structure, biosynthesis and oligogalacturonide-related signalling. Phytochemistry. 2001;**57**:929-967. DOI: 10.1016/S0031-9422(01)00113-3

[36] O'Neill MA, York WS. The composition and structure of plant primary cell walls. In: Rose JKC, editor. The Plant Cell Wall. Boca Raton, Fla.: Blackwell/CRC; 2003. pp. 1-54. DOI: 10.1093/aob/mch185

[37] Yapo BM. Rhamnogalacturonan-I: A structurally puzzling and functionally versatile polysaccharide from plant cell walls and mucilages. Polymer Reviews. 2011;**51**:391-413. DOI: 10.1080/15583724.2011.615 962

[38] Coenen GJ, Bakx EJ, Verhoef RP, Schols HA, Voragen AGJ. Identification of the connecting linkage between homo- or xylogalacturonan and rhamnogalacturonan type I. Carbohydrate Polymers. 2007;**70**:224-235. DOI: 10.1016/j.carbpol.2007.04.007

[39] Seymour GB, Knox JP. Pectins and their Manipulation. Oxford: Blackwell Publishing Ltd; 2002

[40] Yapo BM. Pectic substances: From simple pectic polysaccharides to complex pectins—A new hypothetical model. Carbohydrate Polymers. 2011b;**86**:373-385. DOI: 10.1016/j.carbpol.2011.05.065

[41] Alistair MS, Glyn OP, Peter AW. Food Polysaccharides and their Applications. Florida: CRC Press; 2014. pp. 357-364

[42] Funami T, Zhang G, Hiroe M, Noda S, Nakauma M, Asai I, et al. Effects of the proteinaceous moiety on the emulsifying properties of sugar beet pectin. Food Hydrocolloids. 2007;**21**:1319-1329. DOI: 10.1016/j.foodhyd.2006.10.009

[43] Gullon B. Gómez B, Martínez-Sabajanes M, Yáñez R, Parajó JC, Alonso JL. Pectic oligosaccharides: manufacture and functional properties.

Trends in Food Science and Technology. 2013;**30**:153-161. DOI: 10.1016/j.tifs.2013.01.006

[44] Mohnen D. Pectin structure and biosynthesis. Current Opinion in Plant Biology. 2008;**11**:266-277. DOI: 10.1016/j.pbi.2008.03.006

[45] de Vries JA, Hansen M, Soderberg J, Glahn PE, Pederson JK. Distribution of methoxyl groups in pectins. Carbohydrate Polymers. 1986;**6**:165. DOI: 10.1016/0144-8617(86)90017-2

[46] Leroux J, Langendorff V, Schick G, Vaishnav V, Mazoyer J. Emulsion stabilizing properties of pectin. Food Hydrocolloids. 2003;**17**:455-462. DOI: 10.1016/S0268-005X(03)00027-4

[47] Willats WGT, McCartney L, Mackie W, Knox JP. Pectin: Cell biology and prospects for functional analysis. Plant Molecular Biology. 2001;**47**:9-27. DOI: 10.1023/A:101066291

[48] ED Ngouémazong, Jolie RP, Cardinaels R, Fraeye I, VanLoey A, Moldenaers P, et al. Stiffness of Ca^{2+}pectin gels: Combined effects of degree and pattern of methyl esterification for various Ca^{2+} concentrations. Carbohydrate Research. 2012;**348**:69-76. DOI: 10.1016/j.carres.2011.11.011

[49] Georget D, Ng A, Smith A, Waldron K. The thermo mechanical properties of carrot cell wall material. Journal of the Science of Food and Agriculture. 1998;**78**:73-80. DOI: 10.1016/S0040-6031(98)00276-7

[50] Sila DN, Doungla E, Smout C, Van Loey A, Hendrickx M. Pectin fractions inter conversions: Insight into understanding texture evolution of thermally processed carrot. Journal of Agricultural and Food Chemistry. 2006;**54**(22):8471-8479. DOI: 10.1021/jf0613379

[51] Frank K, Köhler K, Schuchmann HP. Formulation of labile hydrophilic ingredients in multiple emulsions: Influence of the formulation's composition on the emulsion's stability and on the stability of entrapped bioactives. Journal of Dispersion Science and Technology. 2011;**32**:1753-1758. DOI: 10.1080/01932691 .2011.616147

[52] Mcclements DJ, Decker EA, Weiss J. Emulsion-based delivery systems for lipophilic bioactive components. Journal of Food Science. 2007;**72**:R109-R124. DOI: 10.1111/j.1750-3841.2007.00507.x

[53] Nakamura A, Takahash T, Yoshida R, Maeda H, Corredig M. Emulsifying properties of soybean soluble polysaccharide. Food Hydrocolloids. 2004;**18**:795-803. DOI: 10.1016/j.foodhyd.2003.12.005

[54] Surh J, Decker E, McClements D. Influence of pH and pectin type on properties and stability of sodium-caseinate stabilized oil-in-water emulsions. Food Hydrocolloids. 2006;**20**:607-618. DOI: 10.1016/j.foodhyd.2005.07.004

[55] Chassaing K, Koren O, Goodrich JK, Poole AC, Srinivasan S, Ley RE, et al. Dietary emulsifiers impact the mouse gut microbiota promoting colitis and metabolic syndrome. Nature. 2015, 2015;**519**(7541):92-96. DOI: 10.1038/nature14232

[56] Viennois E, Merlin D, Gewirtz A, Chassaing B. Dietary emulsifier-induced low-grade inflammation promotes colon carcinogenesis. Cancer Research. 2017;**77**(1):27-40. DOI: 10.1158/0008-5472.CAN-16-1359

[57] Dickinson E. Hydrocolloids at interfaces and the influence on the properties of dispersed systems. Food Hydrocolloids. 2003;**17**:25-39. DOI: 10.1016/S0268-005X(01)00120-5

[58] Thakur BR, Singh RK, Handa AK. Chemistry and uses of pectin - a review. Critical Reviews in Food Science

and Nutrition. 1997;**37**:47-73. DOI: 10.1080/10408399709527767

[59] Akhtar K, Dickinson E, Mazoyer J, Langendorff V. Emulsion stabilizing properties of depolymerized pectin. Food Hydrocolloids. 2002;**16**:249-256. DOI: 10.1016/S0268-005X(01)00095-9

[60] Dickinson E. Hydrocolloids as emulsifiers and emulsion stabilizers. Food Hydrocolloids. 2009;**23**:1473-1482. DOI: 10.1016/j.foodhyd.2008.08.005

[61] Jung J, Wicker L. Laccase-mediated conjugation of sugar beet pectin and the effect on emulsion stability. Food Hydrocolloids. 2012;**28**:168-173. DOI: 10.1016/j.foodhyd.2011.12.021

[62] Siew CK, Williams PA, Cui SW, Wang Q. Characterization of the surface-active components of sugar beet pectin and the hydrodynamic thickness of the adsorbed pectin layer. Journal of Agricultural and Food Chemistry. 2008;**56**:8111-8120. DOI: 10.1021/jf801588a

[63] Funami T, Nakauma M, Ishihara S, Tanaka R, Inoue T, Phillips GO. Structural modifications of sugar beet pectin and the relationship of structure to functionality. Food Hydrocolloids. 2011;**25**:221-229. DOI: 10.1016/j.foodhyd.2009.11.017

[64] McClements DJ. Protein-stabilized emulsions. Current Opinion in Colloid & Interface Science. 2004;**9**:305-313. DOI: 10.1016/j.cocis.2004.09.003

[65] Yoo S-H, Fishman ML, Hotchkiss AT Jr, Lee HG. Viscometric behavior of high-methoxy and low-methoxy pectin solutions. Food Hydrocolloids. 2006;**20**:62-67. DOI: 10.1016/ j.foodhyd.2005.03.003

[66] Lutz R, Aserin A, Wicker L, Garti N. Structure and physical properties of pectins with block-wise distribution of carboxylic acid groups. Food Hydrocolloids. 2009;**23**:786-794. DOI: 10.1016/j.foodhyd.2008.04.009

[67] Williams PA, Sayers C, Viebke C, Senan C, Mazoyer J, Boulenguer P. Elucidation of the emulsification properties of sugar beet pectin. Journal of Agricultural and Food Chemistry. 2005;**53**:3592-3597. DOI: 10.1021/jf0404142

[68] Castellani O, Al-Assaf S, Axelos M, Phillips GO, Anton M. Hydrocolloids with emulsifying capacity. Part 2- adsorption properties at the n-hexadecane-water interface. Food Hydrocolloids. 2010;**24**:121-130. DOI: 10.1016/j.foodhyd.2009.07.006

[69] Nakai S. Structure-function relationships of food proteins with an emphasis on the importance of protein hydrophobicity. Journal of Agricultural and Food Chemistry. 1983;**31**:676-683. DOI: 10.1021/jf00118a001

[70] Lobato-Calleros C, Recillas-Mota MT, Espinosa-Solares T, Alvarez-Ramirez J, Vernon-Carter EJ. Microstructural and rheological properties of low-fat stirred yoghurts made with skim milk and multiple emulsions. Journal of Texture Studies. 2009;**40**:657-675. DOI: 10.1111/j.1745-4603.2009.00204.x

[71] Liu H, Xu XM, Guo SD. Comparison of full-fat and low-fat cheese analogues with or without pectin gel through microstructure, texture, rheology, thermal and sensory analysis. International Journal of Food Science and Technology. 2008;**43**:1581-1592. DOI: 10.1111/j.1365-2621.2007.01616.x

[72] Lobato-Calleros C, Sosa-Pérez A, Rodríguez-Tafoya J, Sandoval-Castilla O, Pérez-Alonso C, Vernon-Carter EJ. Structural and textural characteristics of reduced-fat cheese-like products made from W1/O/W2 emulsions and skim milk. LWT- Food Science and

Technology. 2008;**41**:1847-1856. DOI: 10.1016/j.lwt.2008.01.006

[73] Pedersen A, Christian H: No and low fat mayonnaise compositions. United States of American Patent No. 5641533; 1997

[74] Endress H-U, Mattes F, Norz K. Pectins. In: Hui YH, editor. Handbook of Science, Technology and Engineering. London: CRC Taylor & Francis; 2006. pp. 1401-1435

[75] Liu H, Xu XM, Guo SD. Rheological, texture and sensory properties of low-fat mayonnaise with different fat mimetics. Food Science and Technology. 2007;**40**:946-954. DOI: 10.1016/j.lwt.2006.11.007

[76] Yashodhara BM, Umakanth S, Pappachan JM, Bhat SK, Kamath R, Choo BH. Omega-3 fatty acids: A comprehensive review of their role in health and disease. Postgraduate Medical Journal. 2009;**85**:84-90. DOI: 10.1136/pgmj.2008.073338

[77] Méndez-Zamora G, García-Macías JA, Santellano-Estrada E, Chávez Martínez A, Durán-Meléndez LA, Silva-Vázquez R, et al. Fat reduction in the formulation of frankfurter sausages using inulin and pectin. Food Science and Technology (Campinas). 2015;**35**:25-31. DOI: 10.1590/1678-457X.6417

[78] Silva-Vazquez R, Flores-Giron E, Quintero-Ramos A, Hume ME, Mendez-Zamora G. Effect of inulin and pectin on physicochemical characteristics and emulsion stability of meat batters. Cyta – Journal of Food. 2018;**16**:306-310. DOI: 10.1080/19476337.2017.1403490

[79] Löfgren C, Guillotin S, Evenbratt H, Schols HA, Hermansson AM. Effects of calcium, pH, and blockiness on kinetic rheological behavior and microstructure of HM pectin gels. Biomacromolecules. 2005;**6**(2):646-652. DOI: 10.1021/bm049619+

[80] Williams PA. Renewable Resources for Functional Polymers and Biomaterials: Polysaccharides, Proteins and Polyesters. Cambridge, UK: Royal Society of Chemistry; 2011. DOI: 10.1039/9781849733519

[81] McClements DJ. Food Emulsions: Principles, Practice, and Techniques. Boca Raton, USA: CRC Press; 2016. https://doi.org/10.1201/b18868

[82] Neirynck N, Van der Meeren P, Gorbe BS, Dierckx S, Dewettinck K. Improved emulsion stabilizing properties of whey protein isolate by conjugation with pectins. Food Hydrocolloids. 2004;**18**:949-957. DOI: 10.1016/j.foodhyd.2004.03.004

[83] Kjøniksen AL, Hiorth M, Nyström B. Association under shear flow in aqueous solutions of pectin. European Polymer Journal. 2005;**41**:761. DOI: 10.1016/j.eurpolymj.2004.11.006

[84] Kuuva T, Lantto R, Reinikainen T, Buchert J, Autio K. Rheological properties of laccase-induced sugar beet pectin gels. Food Hydrocolloids. 2003;**17**:679-684. DOI: 10.1016 / S0268-005X(03)00034-1

[85] Acosta O. Víquez F, Cubero E: Optimisation of low-calorie mixed fruit jelly by response surface methodology. Food Quality and Preference. 2008;**19**(1):79-85. DOI: 10.1016/j.foodqual.2007.06.010

[86] Valdés A, Burgos N, Jiménez A, Garrigós MC. Natural pectin polysaccharides as edible coatings. Coatings. 2015;**5**(4):865-886. DOI: 10.3390/coatings5040865

[87] Bourtoom T. Edible films and coatings: Characteristics and properties.

International Food Research Journal. 2008;**15**:237-248

[88] Ciolacu L, Nicolau AI, Hoorfar J. Global Safety of Fresh Produce. A Handbook of Best Practice, Innovative Commercial Solutions and Case Studies. Sawston, UK: Woodhead Publishing Limited; 2014

[89] Zhang Y, Rempel C, McLaren D. Edible coating and film materials: Carbohydrates. In: Innovations in Food Packaging. 2nd ed. Amsterdam, The Netherlands: Elsevier; 2014. pp. 305-323. DOI: 10.1016/B978-0-12-394601-0.00012-6

[90] Maftoonazad N, Ramaswamy HS. Effect of pectin-based coating on the kinetics of quality change associated with stored avocados. Journal of Food Processing & Preservation. 2008;**32**:621-643. DOI: 10.1111/j.1745-4549.2008.00203.x

[91] Oms-Oliu G, Soliva-Fortuny R, Martín-Belloso O. Using polysaccharide-based edible coatings to enhance quality and antioxidant properties of fresh-cut melon. LWT - Food Science and Technology. 2008;**41**:1862-1870. DOI: 10.1016/j.lwt.2008.01.007

[92] Moalemiyan M, Ramaswamy HS, Maftoonazad N. Pectin-based edible coating for shelf-life extension of ataulfo mango. Journal of Food Process Engineering. 2012;**35**:572-600. DOI: 10.1111/j.1745-4530.2010.00609.x

[93] Menezes J, Athmaselvi KA. Polysaccharide based edible coating on sapota fruit. International Agrophysics. 2016;**30**:551-557. DOI: 10.1515/intag-2016-0019

[94] Menezes J, Athmaselvi KA. Study on effect of pectin based edible coating on the shelf life of sapota fruits. Biosciences, Biotechnology Research Asia. 2016;**13**(2). DOI: 10.13005/bbra/2152

[95] Bayarri M, Oulahal N, Degraeve P, Gharsallaoui A. Properties of lysozyme/low methoxyl (LM) pectin complexes for antimicrobial edible food packaging. Journal of Food Engineering. 2014;**131**:18-25. DOI: 10.1016/j.jfoodeng.2014.01.013

[96] Guerreiro AC, Gago CML, Faleiro ML, Miguel MGC, Antunes MDC. The use of polysaccharide-based edible coatings enriched with essential oils to improve shelf-life of strawberries. Postharvest Biology and Technology. 2015;**110**:51-60. DOI: 10.1016/j.postharvbio.2015.06.019

[97] Abdi S, Roein Z, Erfanimoghadam J, Aziznia S. Application of pectin coating containing essential oil for increasing quality of strawberry fruit. Journal of Postharvest Technology. 2017;**5**(4):83-94

[98] Sanchís E, Ghidelli C, Sheth CC, Mateos M, Palou L, Pérez-Gago MB. Integration of antimicrobial pectin-based edible coating and active modified atmosphere packaging to preserve the quality and microbial safety of fresh-cut persimmon (*Diospyros kaki Thunb. cv. Rojo Brillante*). Journal of the Science of Food and Agriculture. 2017;**97**(1):252-260. DOI: 10.1002/jsfa.7722

[99] Oliveira SM, Brandão TRS, Silva CLM. Influence of drying processes and pretreatments on nutritional and bioactive characteristics of dried vegetables: A review. Food Engineering Reviews. 2015;**8**:134163. DOI: 10.1007/s1239 3-015-9124-0

[100] Silva KS, Garcia CC, Amado LR, Mauro MA. Effects of edible coatings on convective drying and characteristics of the dried pineapple. Food and Bioprocess Technology. 2015;**8**:1465-1475. DOI: 10.1007/s11947-015-1495-y

[101] Canizares D, Mauro MA. Enhancement of quality and stability of dried papaya by pectin-based

coatings as air-drying pretreatment. Food and Bioprocess Technology. 2015;**8**:1187-1197

[102] Pinthus EJ, Saguy IS. Initial interfacial tension and oil uptake by deep-fat fried foods. Journal of Food Science. 1994;**59**:804, 823-807. DOI: 10.1111/j.1365-2621.1994.tb08132.x

[103] Daraei Garmakhany A, Mirzaei HO, Kashani Nejad M, Maghsudlo Y. Study of oil uptake and some quality attributes of potato chips affected by hydrocolloids. European Journal of Lipid Science and Technology. 2008;**110**:1045-1049. DOI: 10.1002/ejlt.200700255

[104] Khalil AH. Quality of french fried potatoes as influenced by coating with hydrocolloids. Food Chemistry. 1999;**66**:201-206. DOI: 10.1016/S0308-8146(99)00045-X

[105] Kizito KF, Abdel-Aal MH, Ragab MH, Youssef MM. Quality attributes of French fries as affected by different coatings, frozen storage and frying conditions. Journal of Agricultural Science and Botany. 2017;**1**(1):23-29

Chapter 5

Pectins as Emulsifying Agent on the Preparation, Characterization, and Photocatalysis of Nano-$LaCrO_3$

Rudy Tahan Mangapul Situmeang

Abstract

The use of environmentally friendly chemicals as the emulsifying agent in the preparation of the advanced materials is a focus and is very interesting to do. Although the focus is important, the advanced material that is made remains a top priority regarding characterization and its activity. One of the chemicals for making advanced and environmentally friendly materials such as $LaCrO_3$ perovskite is pectin functioning as the emulsifying agent. In general, perovskite compounds are materials with very wide applications such as fuel cells, electronic equipment, sensors, magnetic materials, photoluminescent materials, thermic catalysts, and photocatalysts. Furthermore, the use of various solvents to produce perovskite compounds with the aim of getting good applications has been done a lot such as water, alcohols, or other organic solvents, respectively, in mixing precursors directly, precipitation and coprecipitation, microwave, auto-combustion, sol-gel, and hydrothermal methods. In this chapter, preparation of $LaCrO_3$ using pectin as an emulsifying agent will be discussed in advance together with characterization and application of $LaCrO_3$ in the photocatalytic reaction of dyes and cellulose conversion.

Keywords: nanoperovskite, pectin, photocatalysis, dyes, cellulose

1. Introduction

1.1 General introduction of pectin

Pectin is a natural organic compound that has unique structure and characteristics. The uniqueness of the pectin structure is seen from its constituents consisting of three components [1–4], namely, homogalacturonan (HG), rhamnogalacturonan-I (RGI), and rhamnogalacturonan-II (RGII). In principle, the main structure is RG. Furthermore, the functional groups possessed by pectin are ferulic acid, methoxy, acetyl esters, and esters. Interactions between the functional groups that are held provide uniqueness to the characteristics of pectin through hydrogen bonds, hydrophobic interactions, polyelectrolyte behavior, specific ion interactions, and even covalent bonds [5, 6] as external cation binding matrices prepared in the preparation of the advanced material.

Schematically the binding and distribution of cations dissolved in the pectin solution can be illustrated as in **Figure 1**.

In general, pectin can bind various cations with oxidation numbers of +1, +2, and +3 through various functional groups that belong to one or more pectin molecules, so that the cations bound are ready to react to produce a compound that will be well-distributed and the size of the particles produced can reach the nanoscale.

1.2 General introduction of $LaCrO_3$

Perovskite compounds, ABO_3 (where A = cation of alkali, alkaline earth, or lanthanide metal and B = cation of transition metal), have unique chemical and physical properties such as oxidative, magnetic, conductive, refractive, luminescent, and catalytic. With such interesting and valuable characteristics, these compounds have been utilized tremendously in electronic devices as a tuner of the dielectric/ferroelectric responses [7] and an overcomer inefficiency on photovoltaics and other optoelectronic devices [8], sensors as a hydrazine detector [9] and ozone sensing property [10], magnetisms as a huge magnetoresistance [11] and a magnetoelectric response [12], photoluminescences as a highly photoluminescent thin film [13] and a light-emitting material [14], catalysts as CO2/H2 converter into alcohol [15] and pollutant decomposer [16], solid oxide fuel cells as a self-anode in the next power generator [17] and a good performance cathode [18], and photocatalysts as a photo-oxidator of benzylic alcohol [19] and a decomposer of dyes [20].

Perovskite structural material (ABO_3) can be synthesized by mixing the oxide of lanthanide or third main group elements with the oxide of the transition elements. The cations can fit into both the A and B sites of the perovskite structure. In principle, the ABO_3 structure should obey the formulae of Goldschmidt's tolerance factor [21], $t = 0.71(r_A + r_O)/(r_B + r_O)$. This t-value led to the formation of crystalline structures such as cubic [22], orthorhombic [23], and hexagonal forms [24].

One of the perovskite materials, lanthanum chromites ($LaCrO_3$), has been extensively examined due to its applicability as interconnector for solid oxide fuel cell [25], excellent chemical stabilizer [26], good electrical properties at high temperature [27], total oxidation catalyst [28], partial oxidation catalyst [29], oxidative dehydrogenation catalyst [30], and photocatalyst [31]. Nowadays, various kinds of methods have been utilized to prepare the perovskite compounds such as hydrothermal [32, 33], precipitation [34, 35], coprecipitation [36, 37], auto-combustion [38, 39], and sol-gel [40–42]. Among these various kinds of the preparation methods, sol-gel holds particular importance, since it offers many advantages over the others. In the sol-gel technique, a homogeneous product is affected by all steps of preparation such as selection and dissolution of raw material, homogeneous mixing, and pH and temperature adjustments. In addition, these steps led to the opportunity to gain the nanomaterial. Therefore, unique

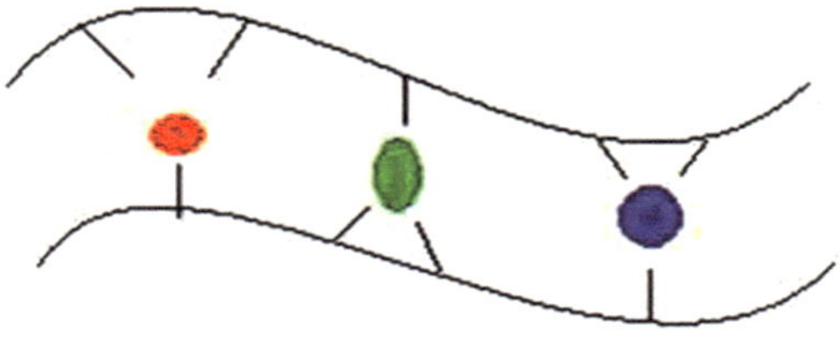

Figure 1.
Illustration of binding of cations by pectin during the dissolution process.

physical and chemical properties of nanomaterial will be more feasible to obtain compared to their micro- and macro-size counterparts in a huge range of updated technologies [43–45].

Since the transformation of raw materials from the dissolved state into a solid state is also crucial in the sol-gel method, gelation, as well as solidification, determines the particle size of the product. To gain the nanomaterial, the agglomeration should be avoided by controlling carefully the process of both gelation and solidification as well as that of thermal treatment.

1.3 Typical property of photocatalyst

In heterogeneous photocatalytic processes, the semiconductors used are chalcogenide-type semiconductor materials (oxides, TiO_2, ZnO, ZrO, and CeO_2, or sulfides, ZnS and CdS). Semiconductors can be used as photocatalyst because they have a void energy region called band-gap energy, which lies between the conduction band boundary (LUMO) and the valence band (HOMO) that does not provide energy for promoting recombination of electrons and holes produced by a photoactivation in these semiconductors.

This semiconductor will function as a catalyst if it is illuminated with photons that have energy that is equal to or more than the energy bandgap (E_g) of the semiconductor used ($hv \geq E_g$). Induction by these rays will excite the electrons (from the valence to the conduction band) in semiconductor materials [46]. As a result of the photon illumination, the formation of electron pairs (e^-) and holes (h^+) which are separated into free photoelectrons in the conduction band and photo hole in the valence band is ready to trigger the reaction as shown in **Figure 2**.

The reaction that occurs in this event is:

$$\text{Semiconductor} + h\upsilon \rightarrow \left(e_{CB}^{-} + h_{VB}^{+}\right)$$

There are several possibilities that occur in electron-hole pairs, namely:

1. Some pairs recombine in particles (volume recombination).
2. Electron-hole pairs recombine on the surface (surface recombination) or in bulk particles in just a few nanoseconds (energy is lost as heat).

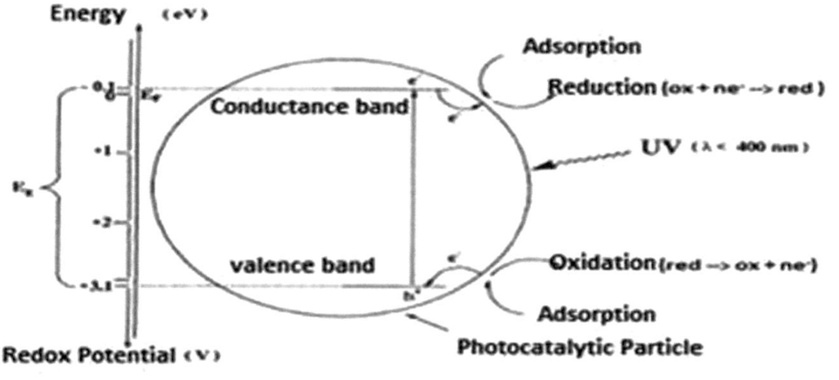

Figure 2.
Band-gap energy diagram in the photocatalytic process ([47]

The electron-hole pair recombination reaction can be written as follows:

$$\text{Semiconductor}\left(e^{-}_{CB} + h^{+}_{VB}\right) \rightarrow \text{semiconductor} + \text{heat}$$

3. Each electron pair can react with donor species (D) and acceptor (A), which are adsorbed on the particle surface. In other words, the electrons in the conduction band that reach the surface will reduce the substrate (A) or solvent on the surface of the particles, while the holes in the valence band will oxidize the substrate (D) either directly or indirectly through hydroxyl radical formation. This phenomenon follows the reaction equation as follows:

$$h\upsilon + \text{semiconductor} \rightarrow e^{-} + h^{+}$$

$$A_{(ads)} + e^{-} \rightarrow A^{-}_{(ads)}$$

$$D_{(ads)} + h^{+} \rightarrow D^{+}_{(ads)}$$

Some possible reactions that can occur with radical ions formed (A^- and D^+) include:

a. A^- and D^+ react between fellow radical ions or react with adsorbates (species adsorbed to the surface).

b. A^- and D^+ combine by transferring the electron back to form an excited state from one of the reactants or releasing heat.

c. A^- and D^+ diffuse from the surface of the semiconductor and participate in chemical reactions that occur in the solution medium.

In general, the process of the occurrence of a photocatalytic reaction based on the energy-gap concept indicates the difference in HOMO energy (the top band of valence contains electrons) and LUMO (the lowest band of conduction without electrons) that must be passed. In other words, the promotion of electrons from the top band of valence to the lowest band of conduction requires minimum energy equivalent to its band-gap energies. If the energy owned is zero or larger than 4 eV, owned are zero or large (>4 eV), then each is a metal or insulator, whereas semiconductor has energy between these values. Furthermore, the band-gap energy is classified as direct and indirect. Direct means that the minimum energy from the lowest band of conduction is just above the maximum energy of the valence band at the same momentum of crystals. If the condition is not so, it is called indirect band-gap energy. The range of band-gap energy possessed by a material will determine the type of energy that will be used so that a photocatalytic reaction can occur. The type of energy can be used for the reaction as shown in **Figure 3**.

In principle, all light electromagnetic waves can be used as an energy source for a chemical reaction process. So far, light that can be used as a trigger for chemical reactions through electron transfer from the HOMO to the LUMO level in the degradation and/or breaking of a compound bond into an environmentally friendly product is visible and ultraviolet.

Visible radiation is often used to degrade toxic compounds of dyes since the waste of dyes which is channeled directly into a river or sea body will have a negative effect

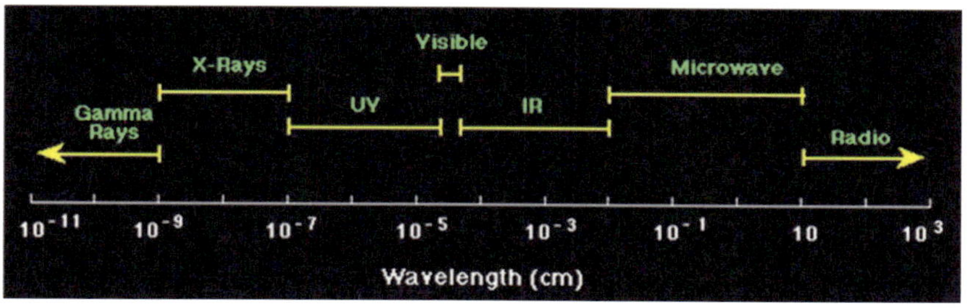

Figure 3.
Schematic radiation energy of electromagnetic wavelength.

Num	Type of bond	Bond energy	
		(kJ/mol)	(eV)
1	H—H	436	11.58
2	C—H	413	10.97
3	O—H	366	9.72
4	C—O	360	9.56
5	C—C	348	9.25
6	C=C	614	16.31
7	C=O	745	19.79
8	C=N	615	16.33
9	C≡C	839	22.29
10	N≡N	941	24.99

Table 1.
Some typical bonds and bond energies.

on aquatic biota. In general, visible light radiation has a wavelength range of 400—800 nm. In other words, the energy needed to break the chemical bonds of a compound is low. Generally, these dyes are compounds that have chromophore groups, such as methine, nitro, azo, anthraquinone, triarylmethane, and phthalocyanine groups. In fact, dyes used in industry can be either natural compounds or syntheses.

Ultraviolet radiation has a high ability to breakdown the bond and cause decomposition because of its high energy compared to infrared radiation and visible light [48]. Sources of ultraviolet radiation can be obtained from sunlight or artificial light. Ultraviolet (UV) radiation of the sun is electromagnetic energy with wavelengths between 200 and 400 nm and has more energy than visible light. Based on its wavelength, solar UV radiation is divided into:

1. UVA with a wavelength of 320—400 nm is a high wavelength and emits radiation of constant magnitude throughout the year. This radiation can cause premature aging of the skin.

2. UVB with a wavelength of 280–320 nm is a shorter wavelength and is more intense than UVA. UVB is more strongly absorbed by several biomolecular pollutants.

3. UVC with a wavelength of 200—280 nm is the most intensive and dangerous UV radiation and has the potential to cause damage to organisms.

Therefore, to choose a type of UV light in photocatalytic reaction depends on the type of bond and the energy required. In general, the type of bond with its energy can be seen in **Table 1**.

2. Method of preparation

Preparation of $LaCrO_3$ catalyst material was carried out using the sol-gel method with pectin as an emulsifying agent. The preparation was conducted by dissolving specified mass of $La(NO_3)_3.9H_2O$ and $Cr(NO_3)_3.6H_2O$, respectively, in 100 mL of pectin solution (4 g pectin). The overall procedure was described in the previous article [49].

3. Characterizations

Before the material made was applied, the photocatalyst was characterized to determine the physical and chemical properties associated through X-ray diffraction analysis, electron transmission microscopy, distribution of particle distribution, energy-gap, and functional groups related to the structural formation and the acidity of Brønsted-Lowry and Lewis.

3.1 Analysis of X-ray diffraction

X-ray diffraction can be utilized to identify the phase formed and the relative percentages of different phases of the materials obtained. Then, the real structural parameters like particle size, lattice parameters (a, b, and c), lattice volume, and theoretical density can be calculated from their diffractogram using Rietveld calculation [50].

X-ray diffraction experiments are carried out by the procedure as described in the previous article [49]. To know the crystallite size, the representative peak of a diffractogram can be elucidated by using the Scherrer method of calculation [51]. The results of the diffractogram, determination of the size of the crystalline phase, and the Rietveld calculation are presented in **Figure 4**.

In general, the results of $LaCrO_3$ prepared using pectin provide a single crystalline phase, nanosize, and other parameter values shown in **Table 2**.

3.2 Transmission electron microscope analysis

TEM can be used to study the morphology and surface characteristics of the perovskite nanomaterials. To evaluate the surface morphology, the samples were

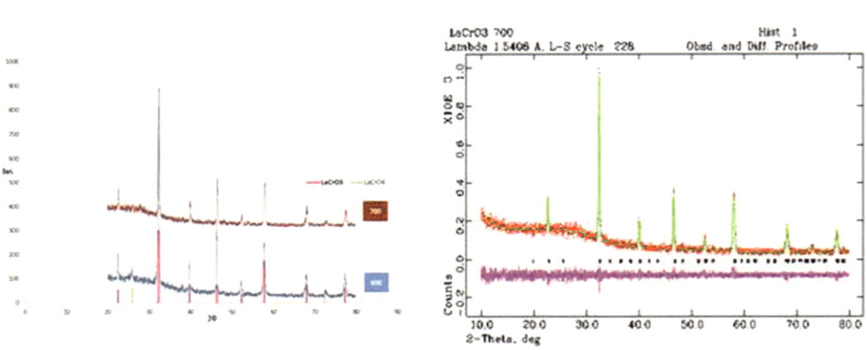

Figure 4.
Diffractogram of $LaCrO_3$ (a) experimental results and (b) the result of the Rietveld calculation using $LaCrO_3$ calcined at 700°C.

Num	Parameter	$LaCrO_3$ calcined at			Ref
		600°C	700°C	800°C	
1	Crystalline phase	Cubic	Orthorhombic	Orthorhombic	[50]
2	Crystalline size (nm)	24.84	24.12	27.09	[48]
3	hkl plane	110	112	121	[50]

Table 2.
Parameters of $LaCrO_3$ prepared in various calcination temperatures.

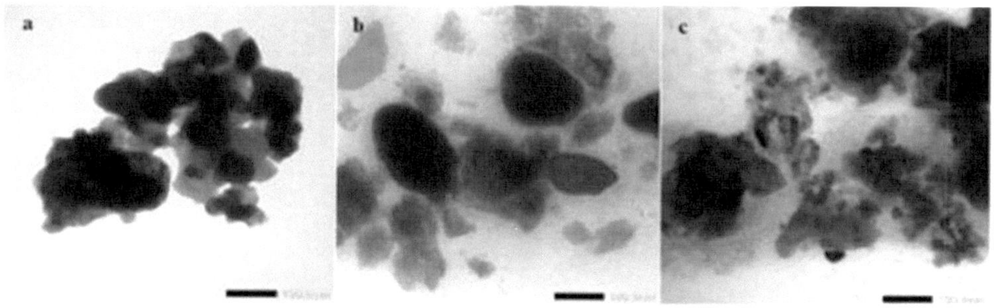

Figure 5.
TEM micrographs of $LaCrO_3$ prepared using pectin as the emulsifying agent: calcined at (a) 600, (b) 700, and (c) 800°C.

characterized using TEM. The analysis was conducted on polished and thermally etched samples with different magnifications. TEM results of $LaCrO_3$ material are presented in **Figure 5**.

It seems that there is still a relatively large region of agglomeration in the crystalline phase formed in each of these preparations. Nevertheless, the particle sizes obtained using TEM in this study are significantly smaller than that of $LaCrO_3$ prepared using sol-gel method reported by another research group [52]. It was found that the particle sizes of the sample calcined at 600, 700, and 800°C are 34.6, 30, and 28 nm, respectively.

3.3 Particle size distribution analysis

Analysis of the particle size distribution of the solid sample was examined by the technique of dynamic light scattering (DLS). The measurement of the sample using this instrument can be determined by either wet or dry method. If a measurement is using the wet method, the sample is prepared using alcohol dispersant such as methanol, ethanol, or propanol. However, if the measurement is using a dry method on preparing the sample, air dispersant could be utilized. More information can be obtained in the manual book [53]. The more important in preparing sample is to prevent the irreversible change to the particle (dissolution, milling, or aggregation) happened.

From **Figure 6**, it can be implied that there are two or three regions of the particle size which are quantum dot, nano-, and micron sizes. Overall, it can be said that nanosize of the particles 21.9, 86.4, and 89.11% referred to $LaCrO_3$ calcined at 600, 700, and 800°C, respectively [49].

The results also can be implied that the more nanosize particle obtained, the higher the temperature of calcination applied. In other words, the temperature of calcination plays a role to determine the nanosize of the particle. In other studies,

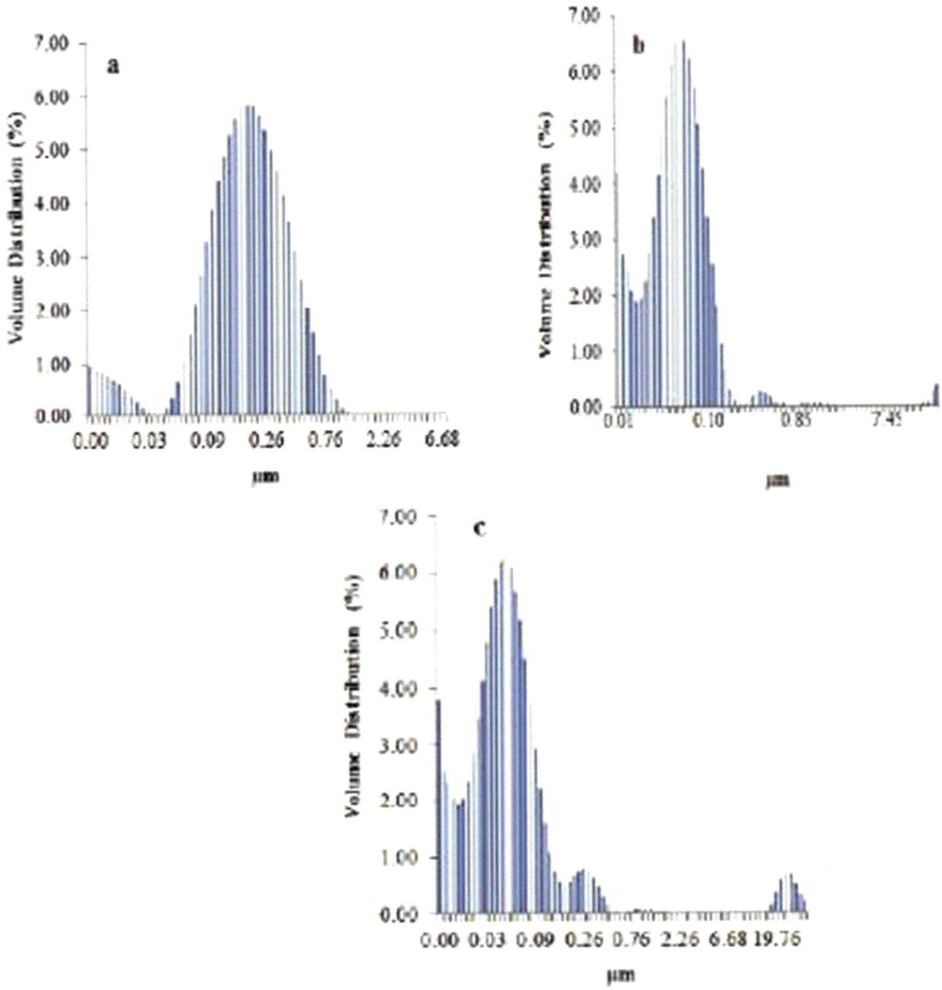

Figure 6.
The particle size distribution of the $LaCrO_3$ calcined at (a) 600, (b) 700, and (c) 800°C.

the particle size of $LaCrO_3$ prepared using hydrothermal method was determined by PSA method. The result proved that the average size of particle is in the range of micron [54] and 57 nm [55].

3.4 Analysis of diffuse reflectance UV-Vis spectroscopy

In order to know the band-gap energy of the $LaCrO_3$ prepared in a different calcined temperature, the analysis was run using diffuse reflectance UV-Vis spectroscopy as shown in **Figure 7**.

To determine the bandgap of a powder sample using the diffuse reflectance spectrophotometer is a common technique [56]. So, in this study the band-gap energy is calculated using a Kubelka-Munk method [57] based on the equation below:

$$\alpha(h\nu) \approx \beta\,(h\nu - E_{op})^{n}$$

where β is a constant and n is an index related to the possible type of electron transition. The value of n could be 1/2, 2, 3/2, and 3. Those values are corresponding

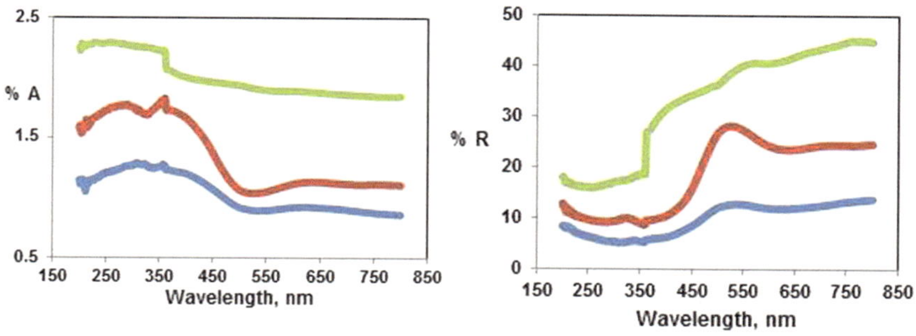

Figure 7.
Reflectance (A) and absorption (B) features of $LaCrO_3$ calcined at 600, 700, and 800°C, respectively.

to the nature of electron transition. In principle, there are two kinds of electron transition, which are direct and indirect. If the n value is 1/2 or 2, it means allowed direct or indirect electron transition happened. But if the n value is 3/2 or 3, it means forbidden direct or indirect electron transition occurred [58].

The results of the band-gap energy from $LaCrO_3$ calcined at 600, 700, and 800°C, respectively, are 2.62, 2.89, and 2.98 eV. The magnitude of those band-gap energies is suitable for photocatalytic reactions using UV and visible light irradiation.

3.5 Fourier transform infrared analysis

In principle, FTIR analysis in the material field is used to assess the functional groups and what bonds are formed in the material prepared in relation to the expected compound. In preparing samples for their analysis, the procedures performed are standard and can be referenced in various libraries [49, 59, 60].

The chemical bonding and chemical structure of the prepared perovskites can be identified. The FTIR spectra can give structural confirmation supporting XRD analysis. Infrared spectra of $LaCrO_3$ material are presented in **Figure 8**.

From **Figure 8**, it can be implied that perovskite $LaCrO_3$ is actually formed and can be assessed based on the type of bond vibration which can be referred to in detail in the following literature [a]. The La—O—La and La—O—Cr bonds through

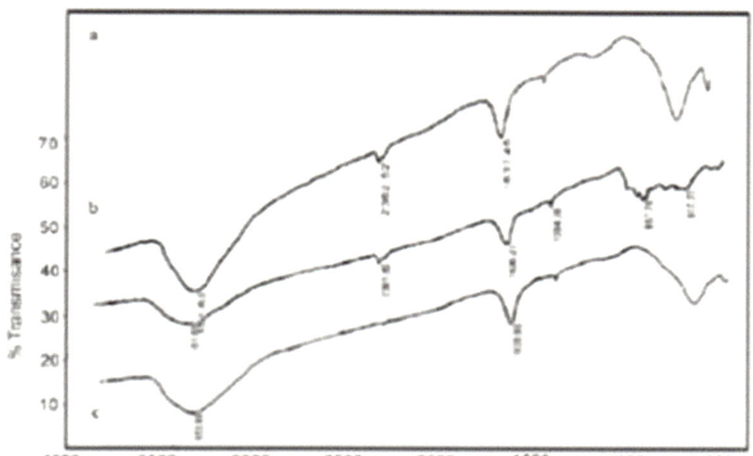

Figure 8.
Infrared spectra of $LaCrO_3$ calcined at (a) 600°C, (b) 700°C, and (c) 800°C [48].

information on bending vibrations are increasingly apparent as the calcination temperature increases. In other words, the structure of $LaCrO_3$ which is formed along with the increase in the calcination temperature is getting closer and can be assessed through the diffractogram data.

The results also reflect that the existence of Brønsted-Lowry and Lewis acid sites is indicated by the presence of absorption bands at wave number 1400 and 1630 cm^{-1}, respectively. In detail, the acidity characteristics of the $LaCrO_3$ calcined at the various temperatures were described in the previous article [49].

4. Applications

The advanced material $LaCrO_3$ prepared using pectin emulsifier and then tested on dye degradation and cellulose conversion was described below.

4.1 Photocatalysis of dye

The photocatalytic activity test on $LaCrO_3$ nanocatalysts was carried out on the methanyl yellow compound by mixing as much as 0.08 g of $LaCrO_3$ nanocatalyst into 300 mL methanyl yellow with a concentration of 100 ppm into a beaker and then homogenized. After that the mixture was pipetted as much as 20 mL with various time variations (0, 10, 20, 30, 40, and 50 min) which had been irradiated by a UV lamp with a distance of 30 cm. After that, UV-Vis spectrophotometry was tested to see the absorbance rate of methanyl yellow.

Then in the photocatalytic reaction for visible light, as much as 0.08 g of $LaCrO_3$ catalyst was put into 300 mL of 100 ppm methanyl yellow in a beaker. Then homogenized it by stirring, while the sample is placed that the mixture is placed under the sun in the range of time 11 am to 1 pm. Then the sample was pipetted as much as 20 mL with various time variations (0, 10, 20, 30, 40, and 50 min) and tested with UV-Vis spectrophotometry to see the absorbance rate of methanyl yellow. Photodegradation under UV light irradiation is shown in **Figure 9**.

4.2 Photocatalysis of cellulose

The application of $LaCrO_3$ photocatalyst to cellulose conversion was carried out on a laboratory scale using procedural as follows:

The $LaCrO_3$ catalyst was then used in the photocatalytic test process for the conversion of nanocellulose to sugar alcohol. As much as 0.5 g of nanocellulose in 100 mL of distilled water are distilled for 30 min. The nanocellulose solution that has been ultrasonified is transferred into a three-neck flask, then stirred with

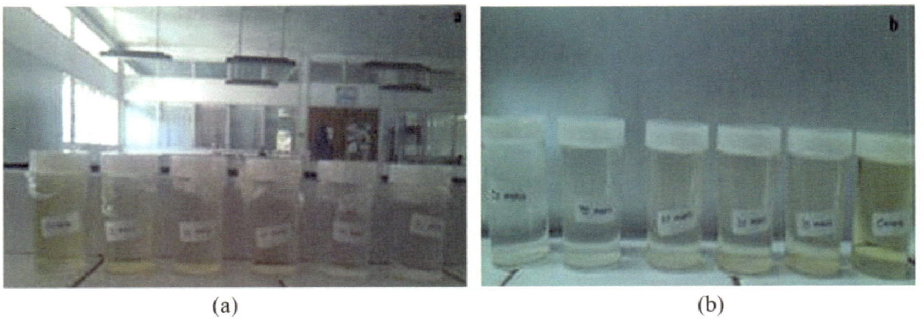

(a) (b)

Figure 9.
Dye photodegradation results under visible light irradiation (a) and ultraviolet (b) for 0–50 min.

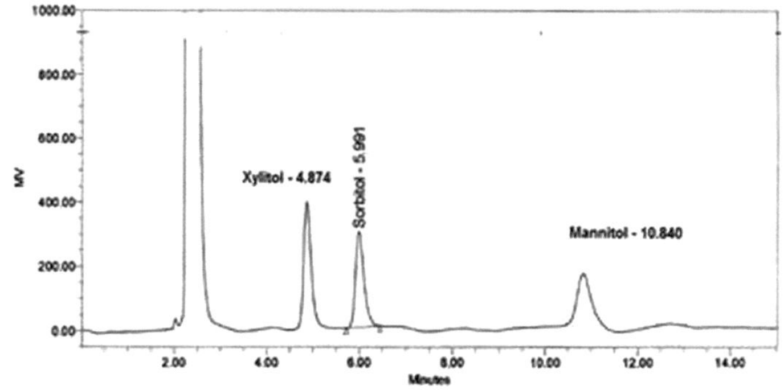

Figure 10.
Photocatalytic cellulose conversion results using $LaCrO_3$ under ultraviolet light irradiation during 1 h exposure.

a stirrer, and then added $LaCrO_3$ nanocatalyst as much as 0.1 g. Furthermore, hydrogen gas is flowed and irradiated with UV light with a variation of time 15, 30, 45, and 60 min. The reaction results were analyzed by high-performance liquid chromatography (HPLC). The results of photocatalytic cellulose conversion using $LaCrO_3$ under UV light are shown in **Figure 10**.

Quantitatively, the results obtained are compared with the standard results for each xylitol sorbitol solution, and the concentration of mannitol is known. Then the quantitative results of the experiment, even though it is still relatively small, are, respectively, 190, 180, and 120 ppm at 1 h exposure with UV radiation [61].

5. Conclusions

Preparation of advanced materials using the sol-gel method with pectin emulsifier is able to make the active catalyst degrade methanyl yellow dyes and convert cellulose into glucose reducing sugars and sugar alcohols such as xylitol, sorbitol, and mannitol. The particle size and crystalline produced by the preparation method can reach nanosize with a range of 24-50 nm. Furthermore, the band-gap energy results state that the $LaCrO_3$ advanced material is in the range of 2.89–3.0 eV.

Acknowledgements

The author would like to acknowledge both the financial support from Indonesian Government through the Directorate Research, Ministry of Research and Higher Education on the contract number 384/UN26.21/PN/2018 in doing the Research and the Research Institution and Community services of the University of Lampung for funding this book chapter.

Author details

Rudy Tahan Mangapul Situmeang
Department of Chemistry, Faculty of Mathematics and Natural Sciences, University of Lampung, Lampung, Indonesia

*Address all correspondence to: rudy.tahan@fmipa.unila.ac.id

References

[1] Willats WG, McCartney L, Mackie W, Knox JP. Pectin: Cell biology and prospects for functional analysis. Plant Molecular Biology. 2001;**47**(1-2):9-27

[2] O'Neill MA, Ishii T, Albersheim P, Darvill AG. Rhamnogalacturonan II: Structure and function of a borate cross-linked cell wall pectic polysaccharide. Annual Review of Plant Biology. 2004;**55**:109-139

[3] Schols HA, Voragen AGJ. Complex pectins: Structure elucidation using enzymes. Progress in Biotechnology. 1996;**14**:3-19

[4] Voragen AGJ, Coenen GJ, Verhoef RP, Schols HA. Pectin, a versatile polysaccharide present in plant cell walls. Structural Chemistry. 2009;**20**:263

[5] Daas PJH, Meyer Hansen K, Schols HA, De Ruiter GA, Voragen AGJ. Investigation of the non-esterified galacturonic acid distribution in pectin with endo-polygalacturonase. Carbohydrate Research. 1999;**318**:135-145

[6] Braccini I, Grasso R, Perez S. Conformational and configurational features of acidic polysaccharides and their interactions with calcium ions: A molecular modeling investigation. Carbohydrate Research. 1999, 1999;**317**(1-4):119-130

[7] Li B-W, Osada M, Kim Y-H, Ebina Y, Akatsuka K, Sasaki T. Atomic layer engineering of high-κ ferroelectricity in 2D perovskites. Journal of the American Chemical Society. 2017;**139**:10868-10874

[8] Tsai H, Asadpour R, Blancon J-C, Stoumpos CC, Ajayan PM, Nie W, et al. Design principles for electronic charge transport in solution-processed vertically stacked 2D perovskite quantum wells. Nature Communications. 2018;**9**:2130

[9] Ali SM, Al Lehaibi HA. Smart perovskite sensors: The electrocatalytic activity of SrPdO for hydrazine oxidation. Journal of The Electrochemical Society. 2018;**165**(9):B345-B350

[10] Kakavelakis G, Petromichelaki V, Gagaoudakis E, Binas V, Kiriakidis G, Petridis K, et al. Solution processed $CH_3NH_3PbI_{3-x}Cl$ perovskite based self-powered ozone sensing element operated at room temperature. ACS Sensors. 2018;**3**:135-142

[11] Borchani SM, Koubaa WC-R, Megdiche M. Structural, magnetic and electrical properties of a new double-perovskite $LaNaMnMoO_6$ material. Royal Society Open Science. 2017;**4**:170920

[12] Casallas F, Vera E, Landinez D, Parra C, Roa J. Structural properties, electric response and magnetic behaviour of $La_2SrFe_2CoO_9$ triple complex perovskite. Journal of Physics Conference Series. 2016;**687**:012047

[13] González-Carrero S, Martínez-Sarti L, Sessolo M, Galian RE, Pérez-Prieto J. Highly photoluminescent, dense solid films from organic-capped $CH_3NH_3PbBr_3$ perovskite colloids. Journal of Physical Chemistry C. 2018;**6**:6771-6777

[14] Lao X, Yang Z, Su Z, Wang Z, Ye H, Wang M, et al. Luminescence and thermal behaviors of free and trapped excitons in cesium lead halide perovskite nanosheets. Nanoscale. 2018;**10**:9949-9956

[15] Situmeang R, Supriyanto R, Septanto M, Simanjuntak W, Sembiring S, Roger AC. $Ni_xCo_yFe_{1-x-y}O_4$ nanocatalyst:

Preparation, characterization, and acitvity in CO_2/H_2 conversion. Proceedings of the 2nd International Conference Indonesian Chemical Society. (October 22-23th, Universitas Islam Indonesia, Yogyakarta) 2013:103-110

[16] Patel F, Patel S. $La_{1-x}Sr_xCoO_3$ (x=0, 0.2) perovskites type catalyst for carbon monoxide emission control from auto-exhaust. Process Engineering. 2013;**51**:324-329

[17] Sengodan S, Ishihara T, Ju Y-W, Shin J, Kwon O, Kim G. Self-decorated MnO nanoparticles on double perovskite solid oxide fuel cell anode by in situ exsolution. ACS Sustainable Chemistry & Engineering. 2017;**5**:9207-9213

[18] Jun A, Kim J, Shin J, Kim G. Perovskite as a cathode material: A review of its role in solid-oxide fuel cell technology. ChemElectroChem. 2016;**3**:511-530

[19] Huang H, Yuan H, Janssen KPF, Solís-Fernández G, Wang Y, Collin Y, et al. Efficient and selective photocatalytic oxidation of benzylic alcohols with hybrid organic-inorganic perovskite materials. ACS Energy Letters. 2018;**3**(4):755-759

[20] Kong J, Yang T, Rui Z, Ji H. Perovskite-based photocatalysts for organic contaminants removal: Current status and future perspectives. Catalysis Today. 2018. DOI: 10.1016/j.cattod.2018.06.045

[21] Goldschmidt VM. Die Gesetze der krystallochimie. Die Naturwissenschaften. 1926;(21):477-485

[22] Wang S, Wu X, Yuan L, Zhang C, Cui X, Lu D. Hydrothermal synthesis, morphology, structure and magnetic properties of perovskite structure $LaCr_{1-x}Mn_xO_3$ (x=0.1, 0.2 and 0.3). CrystEngComm. 2018. DOI: 10.1039/C8CE00421H

[23] Chen Y, Qin H, Shi C, Li L, Hu. High temperature CO2 sensing properties and mechanism of nanocrystalline $LaCrO_3$ with rhombohedral structure: Experiments and ab initio calculations. RSC Advances. 2015;**5**:54710

[24] Coşkun M, Polat O, Coşkun FM, Durmus Z, Çaglar M, Türüt A. The electrical modulus and other dielectric properties by the impedance spectroscopy of $LaCrO_3$ and $LaCr_{0.90}Ir_{0.10}O_3$ perovskites. RSC Advances. 2018;**8**:4634

[25] Kumar Y, Regalado-Pérez E, Ayala AM, Mathews NR, Mathew X. Effect of heat treatment on the electrical properties of perovskite solar cells. Solar Energy Materials & Solar Cells. 2016;**157**:10-17

[26] Rao Y, Wang Z, Chen L, Wu R, Peng R, Lu Y. Structural, electrical, and electrochemical properties of cobalt doped $NiFe_2O_4$ as a potential cathode material for solid oxide fuel cells. International Journal of Hydrogen Energy. 2013;**38**(4):14329

[27] Tiwari B, Dixit A, Naik R, Lawes G, Rao MSR. Dielectric and optical phonon anomalies near antiferromagnetic ordering in $LaCrO_3$: A possible near room temperature magnetodielectric system. Applied Physics Letters. 2013;**103**:152906

[28] Zhang Z, Kong Z, Liu H, Chen Y. Frontiers of Chemical Science and Engineering. 2014;**8**(1):87-94

[29] Khine MSS, Chen L, Zhang S, Lin J, Jiang SP. International Journal of Hydrogen Energy. 2013;**38**:13300-13308

[30] Dinse A, Frank B, Hess C, Habel D, Schomäcker R. Journal of Molecular Catalysis A: Chemical. 2008;**289**:28-37

[31] Kanhere P, Chen Z. Mole. 2014;**19**:19995-20022

[32] Kang M, Yun J, Cho C, Kim C, Tai W. Open Journal of Inorganic Non-metallic Materials. 2013;**3**:37-42

[33] Rivas-Vázquez LP, Rendón-Angeles JC, Rodríguez-Galicia JL, Gutiérrez-Chavarria CA, Zhu KJ, Yanagisawa K. Journal of the European Ceramic Society. 2006;**26**(1-2):81-88

[34] Chettapongsaphan C, Charojrochkul S, Assabumrungrat S, Laosiripojana N. Asian Journal of Energy and Environment. 2008;**9**(1-2):101-119

[35] Doggali P, Grasset F, Cador O, Rayalu S, Teraoka Y, Labhsetwar N. Journal of Environmental Chemical Engineering. 2014;**2**:340

[36] Kumar S, Teraoka Y, Joshi AG, Rayalu S, Labhsetwar N. Journal of Molecular Catalysis A: Chemical. 2011;**348**:42

[37] Yanping W, Junwu Z, Xiaojie S, Xujie Y. Journal of Rare Earths. 2007;**25**:601-610

[38] Corrêa HPS, Paiva-Santos CO, Setz LF, Martinez LG, Mello-Castanho SRH, Orlando MTD. Powder Diffraction. 2008;**23**(2):S18-S22

[39] Tong Y, Ma J, Zhao S, Huo H, Zhang H. Journal of Nanomaterials. 2015;**115**:1-5

[40] Doggali P, Rayalu S, Teraoka Y, Labhsetwar N. Journal of Environmental Chemical Engineering. 2015;**3**:420

[41] Meng M, Guo X, Dai F, Li Q, Zhang Z, Jiang Z, et al. Applied Catalysis B: Environmental. 2013;**142-143**:278-289

[42] Situmeang R, Manurung P, Sulistiyo ST, Hadi S, Simanjuntak W, Sembiring S. Asian Journal of Chemistry. 2015;**27**(3):1138-1142

[43] Girish H-N, Shao GQ , Basavalingu B. RSC Advances. 2016;**6**:79763-79767

[44] Chen W, Chen X, Yang Y, Yuan J, Shangguan W. International Journal of Hydrogen Energy. 2014;**39**:13468-13473

[45] López-Suárez FE, Bueno-López A, Illán-Gómez MJ, Trawczynski J. Applied Catalysis A: General. 2014;**485**:214-221

[46] Richardson JT. Principles of Catalyst Development. New York: Plenum Press; 1989. xv, 288 pp., ISBN 0-306-43162-9

[47] Herrmann J-M. Heterogeneous photocatalysis: Fundamentals and applications to the removal of various types of aqueous pollutants. Catalysis Today. 1999;**53**:115-129

[48] Ibhadon AO, Fitzpatrick P. Heterogeneous photocatalysis: Recent advances and applications. Catalysts. 2013;**3**(1):189-218

[49] Situmeang R, Supryanto R, Kahar LNA, Simanjuntak W, Sembiring S. Characteristics of nanosize $LaCrO_3$ prepared through sol-gel route using pectin as emulsifying agent. Oriental Journal of Chemistry. 2017;**33**(4):1705-1713

[50] Rietveld HM. A profile refinement method for nuclear and magnetic structures. Journal of Applied Crystallography. 1969;**2**:65-71

[51] Cullity BD, Stock SR. Elements of X-Ray Diffraction. 3rd ed. Prentice-Hall Inc.; 2001. pp. 167-171. ISBN 0-201-61091-4

[52] Rativa-Parada W, Gomez-Cuaspud JA, Vera-Lopez E, Carda JB. Structural and electrical study of $LaCrO_3$ modifiedwith Fe and Co. Journal of Physics: Conference Series. 2017;**786**:012029

[53] Rawle A. A Basic Guide to Particle Characterization. Malaysia: Malvern Instrument Limited; 2012. pp. 1-8

[54] Sardar K, Lees Martin R, Kashtiban RJ, Sloan J, Walton Richard I. Direct hydrothermal synthesis and physical properties of rare-earth and yttrium orthochromite perovskites. Chemistry of Materials. 2011;**23**:48-56

[55] Athawale AA, Desai PA. Silver doped lanthanum chromites by microwave combustion method. Ceramics International. 2011;**137**:3037-3043

[56] López R, Gómez R. Band-gap energy estimation from diffuse reflectance measurements on sol–gel and commercial TiO_2: A comparative study. Journal of Sol-Gel Science and Technology. 2012;**61**:1-7

[57] Sangiorgi N, Aversa L, Tatti R, Verucchi R, Sanson A. Spectrophotometric method for optical band gap and electronic transitions determination of semiconductor materials. Optical Materials. 2017;**64**:18-25

[58] Urbach F. The long-wavelength edge of photographic sensitivity and of the electronic absorption of solids. Physical Review. 1953;**92**(5):1324

[59] Yazýcý DT, Bilgiç C. Determining the surface acidic properties of solid catalysts by amine titration using Hammett indicators and FTIR-pyridine adsorption methods. Surface and Interface Analysis. 2010;**42**(6):959-962

[60] ASTM 4824-13. Test method for determination of catalyst acidity by pyridine chemisorption. MNL 58-EB. 2013

[61] Situmeang R, Sembiring Z, Simarmata E, Tamba M, Yuliarni T, Simanjuntak W, Sembiring S. Advances in Natural Sciences: Nanoscience and Nanotechnology. 2019;**10**(1):0150xx

Section 3

Pectin Interactions

Chapter 6

Properties of Wine Polysaccharides

Leticia Martínez-Lapuente, Zenaida Guadalupe and Belén Ayestarán

Abstract

Polysaccharides are the main macromolecules of colloidal nature in wines, and play a fundamental role in the technological properties and organoleptic characteristics of the wines. The role of the different wine polysaccharides will depend on their quantity but also on their chemical composition, molecular structure and origin. Wine polysaccharides originate from grapes and yeast acting during the winemaking. The main polysaccharides present in wines can be grouped into three major families: (i) polysaccharides rich in arabinose and galactose (PRAG), (ii) polysaccharides rich in rhamnogalacturonans (RG-I and RG-II), which both come from the pectocellulosic cell walls of grape berries, and (iii) mannoproteins (MP) released by yeasts. This paper describes the origin, structure and role of the different wine polysaccharide families through a bibliographic revision of their origin and extraction into the wines, as well as their technological and sensory properties.

Keywords: wine, rhamnogalacturonans, polysaccharides rich in arabinose and galactose, rhamnogalacturonans, mannoproteins, technological and sensory properties

1. Introduction

Polysaccharides are the main macromolecules of colloidal nature in wines. Therefore, these compounds play a fundamental role in the technological properties and organoleptic characteristics of the wines.

The content of the different polysaccharide families in the wines depends mainly on the grape variety and its degree of maturation, the winemaking technology used (including type of strain of yeast and bacteria), and the transformation of the polysaccharides during the wine aging process [1–5]. These macromolecules show different technological properties in wines. Wine polysaccharides are widely known for their effect on the physicochemical stabilization of wine; thus, they are able to interact with the colloidal particles present in wines, reducing their reactivity and limiting their aggregation and flocculation [6]. These macromolecules have the ability to interact and aggregate with tannins [7], prevent the formation of protein haze in white wines [8], and delay or even arrest the outgrowth of the crystals of potassium bitartrate to a macroscopic visual size [9]. Wine polysaccharides have also been associated to the mouthfeel perceptions because they are able to modify the sensory properties of wines [7, 10]. Several authors [10, 11] have observed that wine polysaccharides can modulate the astringency perception, increasing the sweetness sensation and body. Astringency is usually defined as the array of tactile sensations felt in the mouth including shrinking, puckering and tightening of the oral surface. In addition, polysaccharides are able to interact with wine volatile compounds [12], and thus affect the aroma of the wines.

Polysaccharides are extracted during the mechanical operations applied to the grapes (destemming-crushing, pressing and pumping of the crushed destemmed grapes) and during some stages of the winemaking. Therefore, polysaccharides are released in white, rosé and red winemaking during the premaceration process before starting the alcoholic fermentation, but also during the maceration fermentation of the red wine elaborations, and during the aging of the wines on their lees. On the contrary, other stages of the winemaking, such as filtration, produce a decrease in the content of wine polysaccharides [5].

Wine polysaccharides come from both the cell walls of the grape itself, and the yeasts and other microorganisms that act during the winemaking process. **Figure 1** shows a classification of the polysaccharides present in wines according to their origin.

From an oenological point of view, polysaccharides from grapes and yeasts are the most important both quantitatively and qualitatively. Therefore, the main polysaccharides present in wines can be grouped into three major families: (i) polysaccharides rich in arabinose and galactose (PRAG) [13] and (ii) polysaccharides rich in rhamnogalacturonans (RG-I and RG-II), which both come from the pectocellulosic cell walls of the grape berries [13], and (iii) mannoproteins (MP) produced and released by yeasts during the fermentation and the aging of wines on their lees [8]. Other wine polysaccharides such as glucans, produced by *Botrytis cinerea*, only become relevant when an infection with this fungus occurs, causing difficult clarifications and filtrations. Bacterial polysaccharides are present in the wines in very low concentrations. Polysaccharides exogenous to wine include carboxymethylcellulose and arabic gum, which are additives allowed by the International Organization of Vine and Wine (OIV).

Among all these types of polysaccharides, not all show the same behavior with respect to wines, and their concrete effects and properties will depend on their size, chemical composition, molecular structure and origin.

The objective of the present paper is to describe the origin, structure and key role of the different wine polysaccharide families through a bibliographic revision of their origin and extraction into the wines, as well as their technological and sensory properties.

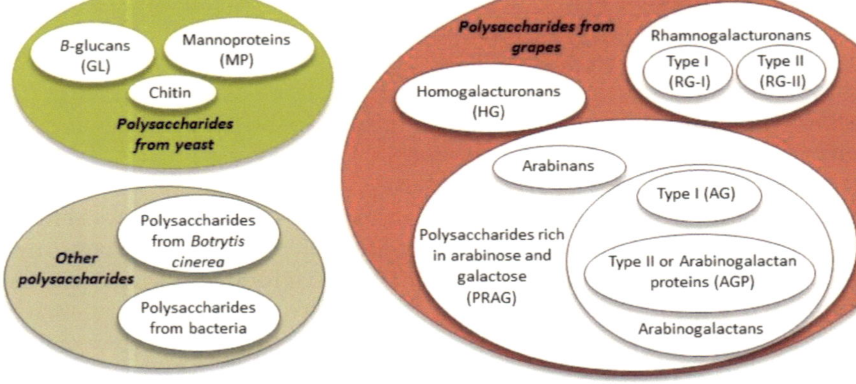

Figure 1.
Classification of the main polysaccharides of wines according to their origin.

2. Grape polysaccharides: origin, structure and functions

The plant cell wall is composed of a highly integrated and structurally complex network of polysaccharides, including celluloses, hemicelluloses and pectins, and also structural proteins [14]. Pectins are a family of heteropolysaccharides characterized by a high content of α-D-galacturonic acid residues partially methyl esterified [15]. These heteropolysaccharides are located in the middle lamella of the primary cell walls, and are mainly composed of a galacturonic acid backbone and chains of several monosaccharides. The smooth region is represented by homogalacturonans (HG), which are galacturonic acid chains more or less methylated/acetylated; the hairy region (high density of side chains) is composed of rhamnogalacturonans type I (RG-I) and type II (RG-II) [16]. RG-I consists of rhamnose and galacturonic acid and represents a very small proportion of grape-based pectins; RG-II is formed in the grape berry during the maturation and is released into the wine during the winemaking. Arabinogalactan proteins (AGP) are glycoproteins also located in the plant cell walls and extracted during the winemaking. They are themselves sidechains of the backbone that arise from the hairy region of pectins and are connected via specific hydroxyproline-rich proteins and, together with arabinogalactans, contribute to the so-called *polysaccharides rich in arabinose and galactose* (PRAG) [17]. Hemicellulose is formed by several polymeric structures in which xyloglucan (a backbone of cellulose with side chains containing xylose, galactose and fucose) is the most abundant [18]. Cellulose microfibrils represent the major constituent of the cell wall polysaccharides, and they are interacting with hemicellulose and pectic polysaccharides, improving the structural integrity of the plant cell wall [19].

Grape berries are composed of three main tissue types [20]: the skins, the pulp and the seeds. The structural properties of the cell walls of grape berries, especially the cell walls from the exocarp (the skin), determine the mechanical resistance, the texture, and the ease of processing berries. Grape skins represent about 5–10% of the total dry weight of the grape berry, and act as a hydrophobic barrier to protect the grapes from physical and climatic injuries, dehydration, fungal infection and UV light. The grape skin itself can be divided into three superimposed layers (**Figure 2**) [21]: (1) the outermost layer, the cuticle, is composed of hydroxylated fatty acids called cutin, and is covered by hydrophobic waxes; (2) the intermediate epidermis, assumed to consist of one or two layers, which appears as a regular tilling of cells; and (3) the inner layer, the hypodermis, which is the layer closest to the pulp, and which is composed of several cell layers that contain most of the phenolics

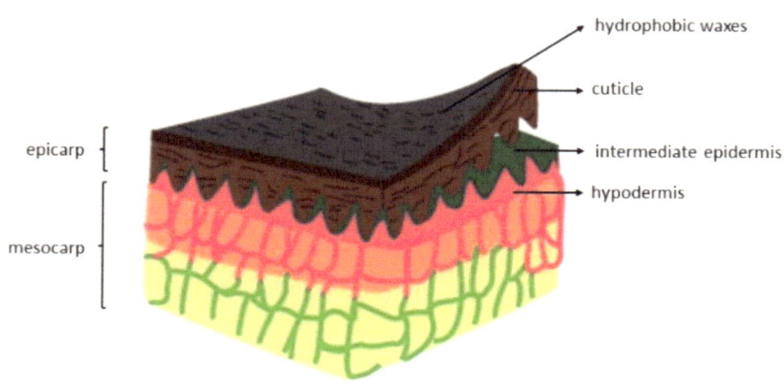

Figure 2.
Different layers of the grape skin.

in grape skin [22]. The cuticle, that covers the skin, is the primary interface between the plant and the environment and is a protective layer (against pathogens and minimizes water loss) that consists of waxes (soluble lipids) embedded in or deposited on the cutin-rich matrix [23]. Gao et al. [24] describe that this wax layer most probably, in red winemaking, albeit not proven, prevents cell wall degrading enzymes from penetrating into the inner tissues (skin and pulp), thus enzymes can only penetrate effectively from the pulp exposed during grape crushing.

The cell walls from the skin form a barrier to the diffusion of components such as aromas and polyphenols, which are important to the quality of the wines. Phenolic compounds contribute to color, astringency and bitterness of the red wines. Aroma is one of the major factors that determine the quality of the wine, showing the skins more than a half of the volatile compounds present in the grape berries [19]. It is well known that the grape berry skin cell walls consist of cellulose, hemicellulose, and are particularly rich in pectin [13, 25]. This pectin component contains a number of polymers HG, RG-I, side chains such as arabinans and galactans, RG-II and AGP [25, 26], and was proposed to be associated with other cell wall polymers (cellulose and hemicellulose) [27].

The pulp (i.e., flesh, also known as pericarp) is the main storage tissue for free sugars (i.e., glucose and fructose) and organic acids (i.e., tartaric acid) [28]. Pulp cells and tissues expand significantly during and after the veraison stage by volume compared to skin cells which expand by net surface area (i.e., a surface-to-volume ratio) [27]. Pulp tissue cell wall layers comprise mainly cellulose and pectin polysaccharides in addition to extension proteins [27].

The ease of skin degradation is directly linked to the skin cell wall composition and morphology [29], and the grape origin [29] and cultivar. Ortega-Regules et al. [30] points out that the differences among morphology and composition of the skin and pulp cell wall of three different red grape varieties (Monastrell, Syrah, Cabernet Sauvignon, Merlot) could explain the different anthocyanin extractability during the winemaking process. Moreover, the liberation of polysaccharides into the wine from the degradation of the grape cell wall could also be affected by an increase of the cell wall rigidity.

Grape berry ripening consists of a cell division (green) phase followed by a cell expansion (ripe) phase [25]. The onset of this second phase known as veraison is marked by the initiation of events such as sugar accumulation, a decrease in organic acids, color development, berry expansion and fruit softening.

The process of ripening, characterized in many fruits by softening of the fleshy tissues, is primarily due to textural changes partially correlated with cell wall polysaccharide remodeling [31]. Berry ripening links with size and morphological changes and a series of coordinated biochemical processes. Both biosynthetic and degradative metabolism of cell wall components involve numerous plant enzymes. Several reviews [32–34] discuss in detail the processes and the enzymes involved in plant cell wall turnover. In grapes, the changes in the cell wall structure involve the solubilization of galacturonan, with a concomitant reduction in the abundance of the arabinogalactan side chains of pectins [35], which can play a role in phenolic extractability [36]. It is thought that the loss of these components opens the interior of the cell wall to several degrading enzymes, causing further depolymerization, and an increased porosity [37]. The progressive pectin degradation of the grape skin cell walls [38] that takes place thorough ripening, should favor polysaccharide solubilization in the juice and thus in wine [39]. Martínez-Lapuente et al. [40] observed that the grape ripening stage (premature and mature grapes) showed a significant impact on the content, composition, and evolution of polysaccharides of sparkling wines. PRAG, RG-II, and oligosaccharides in base wines increased with maturity.

Pectins are among the plant polysaccharides found in wines, and are present in concentrations ranging from around 200 to 1500 mg/L [41]. Polysaccharide amounts depend on different parameters that include the grape variety, terroir, maturity stage, vintage, the wine-making techniques, and the treatments leading to increased solubilization of the macromolecular components of grape berry cell walls [4].

Several researches have studied the effect of techniques and treatments that could increase the solubilization of the polysaccharides of the grape cell walls. Some of them looked for the bursting of the grape cells, thus promoting the breakdown of the linkages stiffening the structure of the grape cell walls and allowing an increase in the release of the polysaccharides. The press fractioning, for example, allowed to segregate the grape juices with different qualities. Jégou et al. [42] observed significant changes in the polysaccharide and oligosaccharide base wine composition and concentration as the pressing cycle of the grapes progressed. The crushed berry is other technique used to physically break the grape berry cell walls, causing depectination and the release of cell wall polysaccharides in significant amounts into the fermenting must [2, 24]. Another technique consists in lowing the temperature of the entire or broken grapes. Low temperature techniques (cold prefermentative maceration, addition of dry ice at the beginning of the fermentation, and grape skin freezing) are additional tools used for degrading the cell wall and achieving greater extraction of polysaccharides [4]. Dry ice addition at the beginning of the fermentation has also proven a significant influence on the polysaccharide concentration and composition of the wines made from a given cultivar, whereas cold prefermentative maceration or grape skin freezing showed no effect [4, 43]. Flash release and heating accelerated the extraction of grape polysaccharides [44]. On the contrary, wines obtained by pressing immediately after flash release contained lower amounts of polyphenols and grape polysaccharides than those made with pomace contact, indicating that the extraction continued during the maceration. Flash release, consisting of the heating of the grapes in a closed tank and then placing them under vacuum, is used to break the cell walls and cool the must.

Other techniques such as modified skin contact times enhanced the release of polysaccharides. Prefermentative maceration at 18°C could also be applied to increase the content of polysaccharides in the wines [3]. The polysaccharides are gradually extracted during the maceration and the alcoholic fermentation due to grape tissue breakdown and degradation of the grape berry cell wall [2, 36]. Polysaccharide concentration increases during skin contact and is much higher in red wines than in white wines [45]. The commercial enzymes have been traditionally used in wine elaborations in order to produce a progressive cell wall disassembly during the winemaking and, hence, improve the release of valuable grape skin compounds such as the anthocyanins [46], aroma components [47], polysaccharides and oligosaccharides [48, 49]. Ayestarán et al. [50] analyzed the influence of commercial enzymes on the wine polysaccharide content, and reported that wines treated with commercial enzymes had higher concentrations of AGP and RG-II than control wines, probably due to the ability of commercial enzymes to hydrolyze the grape pectic polysaccharides during the maceration-fermentation stage. However, contradictory results have been obtained in other studies [48, 51, 52], probably due to the different activities and nature of the commercial preparations. RG-II, containing rare sugars, is also abundant in wines as it resists enzymatic degradation [53].

Guadalupe and Ayestarán [2] studied the changes occurring on the must and wine polysaccharide families of the grape cell walls during the different stages of the red wine processing, including maceration-fermentation and post maceration, malolactic fermentation, and oak aging and bottle aging. Passing from must to wine produced a loss of low-molecular-weight grape structural glucosyl polysaccharides, and an important increase of grape-derived AGP, and RG-II. AGP were more easily extracted tan RG-II, and small quantities of RG-II monomers and galacturonans were detected. Post maceration produced a reduction in all grape polysaccharide families, particularly acute in AGP. The reduction of polysaccharides during malolactic fermentation only affected grape AGP. Wine oak and bottle aging was associated with a relative stability of the polysaccharide families. AGP were thus the majority polysaccharides in young wines. Precipitation of polysaccharides was noticeable during the winemaking, and it mainly affected to the high-molecular-weight AGP. Hydrolytic phenomena affected the balance of wine polysaccharides during late maceration-fermentation. Other authors [3, 54, 55] have observed a change to lower molecular weight polysaccharides during the wine aging, suggesting a partial degradation of the polysaccharides during the aging on lees, and a modification of their properties and solubilization. Pati et al. [56] concluded that the aging on lees led to an increase in all wine polysaccharide glycosyl residues, with the exception of glucose, xylose and myo-inositol, and to volatile profile modifications. The concentrations of cell wall polysaccharides are affected by the filtration process. Therefore, cross-flow microfiltration has shown to produce the highest retention of polysaccharides and proanthocyanidins in all the wines, mainly PRAG and highly polymerized phenols [5]. AGP greatly affected the filtration processes [57].

The final concentrations of cell wall polysaccharides that are extracted during the maceration and alcoholic fermentation are important for wine colloidal stability. RG-II and AGP can enhance or inhibit tannin self-aggregation [7, 58, 59]. Watrelot et al. [60] describe that the main interactions that occur between tannins and polysaccharides are hydrophobic interactions and hydrogen bonds, which differently affect the body, structure and mouthfeel sensations of the wines [10, 61]. Brandão et al. [11] studied the effect of two wine polysaccharides (AGP and RG-II) on the salivary proteins-polyphenol interactions. In general, both polysaccharides were effective to inhibit or reduce salivary proteins-polyphenol interactions and aggregations, and thus both polysaccharides were able to affect the astringency of wines and other beverages and foods. Different researches also point out that AGP show a protective effect against protein haze in white wines [8, 62], while RG-II increases tartrate crystallization at low concentrations and inhibit it at high concentrations [63]. Recent studies suggest that grape AGP do not affect the foamability of sparkling wines but increase foam stability [64, 65].

Aroma compounds can physically or chemically interact with other wine matrix components such as polyphenols, glycoproteins, and polysaccharides. One of the most important factors that can limit the rate of release of aroma compounds during wine consumption could be the interaction between aroma and non-volatile matrix components. This interaction can change the distribution of the aroma compounds between the aqueous solution and the vapor phase (partition coefficient), and thus, alter the odorant volatility, and influence the headspace partitioning of volatiles producing two opposite effects: a retention effect, decreasing the amount of aroma in the headspace, or a "salting out" effect, causing an increase in the headspace concentration of a volatile compound because of the increase in the ionic strength of the solution [66]. Some authors [67, 68] have observed that the addition of arabinogalactan compounds to wines at low concentrations increases the volatility of the aroma compounds.

3. Yeast polysaccharides: origin, structure and functions

Mannoproteins (MP) are polysaccharides released into the wines by *Saccharomyces cerevisiae* yeast either during the fermentation when yeast are actively growing, or after the yeast autolysis by the action of glucanases on the cell wall during aging [69]. The amount of MP released by yeast depends on the specific yeast strain [70] and the winemaking and aging conditions [57]. MP are the second most abundant class of polysaccharides found in wine [2, 13, 42]. It is estimated that MP is around 35% of total polysaccharides in red wines [13], ranging approximately from 100 to 150 mg/L [71].

MP are located in the outermost layer of the yeast cell wall and can account for up to 50% of the cell wall dry mass of *Saccharomyces cerevisiae* [72]. The structure of MP present in wines has been described in several papers [8, 73]; basically, it consists of many small chains with one-to-four D-mannose residues in α-(1 → 2) or (1 → 3), which are linked to polypeptide chains on serine or threonine residues (**Figure 3**).

Wine MP are often highly glycosylated, with carbohydrate fractions consisting mainly of mannose (>90%) and glucose [69], and proteins ranging from 1 to 10% [13, 42, 72]. It has been reported sizes that vary within the range 5–800 kDa [1], with typical range between 50 and 500 kDa [13]. MP can be hydrolyzed by α-mannosidases and proteases, releasing small peptidomannans into the wine [1]. At wine pH, MP carry negative charges and they may establish electrostatic and ionic interactions with other components of the wine [74], resulting in the formation of complexes in a process that is dependent on their net electrical charge and on the structure of their functional groups [75].

MP in wines have great relevance from both a technological and a sensorial point of view [76], although they may be responsible for a decrease in wine color intensity or lower filterability [77, 78]. The different oenological functions of the yeast MP are discussed below.

MP seem to protect wines against protein precipitation. Protein haze is due to the instability of the grape proteins that occur naturally in wines [79], their denaturation and precipitation. It is often related to exposure to high temperatures but can also develop in properly stored wines [80, 81]. Moine-Ledoux and Dubourdieu [82] identified a 32-kDa fragment of *S. cerevisiae* invertase capable of reducing protein haze in white wines, and similar properties were observed for the intact protein [83]. Other yeast cell wall proteins have been shown to stabilize wine against protein haze [84] by reducing protein aggregate particle size [84]. In fact, MP could interact with heat-unfolded proteins, thus preventing protein self-aggregation by limiting the

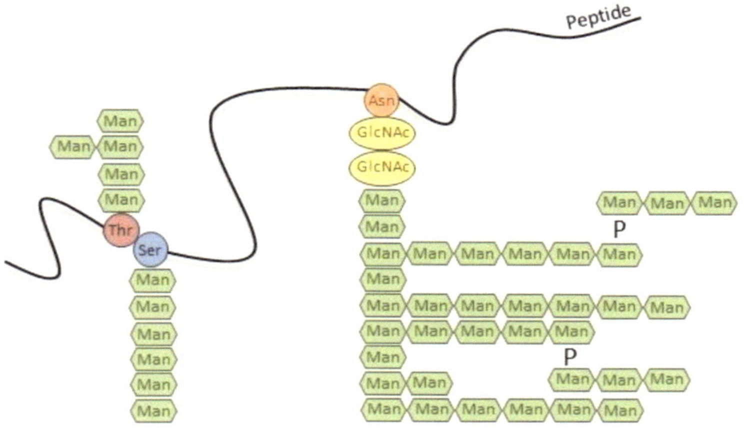

Figure 3.
Chemical structure of yeast exocellular mannoproteins. Asn, asparagine; GNAc, N-acetylglucosamine; man, mannose; P, phosphate; Ser, serine; Thr, threonine.

availability of some protein binding sites with a steric hindrance mechanism [85]. This effect seems to be dependent on the yeast used and the composition and size of the polysaccharides released [86] as well as pH and the ionic strength [87]. However, other authors have revealed that polysaccharides modulate the aggregation kinetics and final haziness, interfering with the aggregation process, but could not prevent it [87]. The ability of a yeast MP to stabilize wine proteins has been attributed specifically to the glycan portion of the proteoglycan [88]. Moreover, protein stabilization effectiveness in white wines has been related to MP chemical composition, concretely with their high mannose to glucose ratio [89].

MP play also an important role in tartrate salt crystallization. Several studies have shown that MP inhibit the crystallization of tartrate salts by lowering the crystallization temperature, particularly sharply glycosylated MP of medium molecular weight (30–50 kDa) [63, 90]. Other authors mention that MP affect the rate of crystal growth by binding to the nucleation points and preventing the expansion of the crystal structure [91]. The mechanism of mannoprotein's impact on tartrate stability is thought to be based on a competitive inhibition, which limits crystal formation [92]. MP act in the first stage of the formation of bitartrate crystals, and also during its growth, preventing the precipitation of the crystals [92]. It is also described that MP do not prevent potassium bitartrate nucleation. Instead, these compounds seem to delay or even arrest the outgrowth of the crystals to a macroscopic, visual size [9]. According to Moine-Ledoux and Dubourdieu [90], the stabilizing effect of MP may delay the appearance of crystals for a month in relation to the untreated wine. It was observed that a dose of 25 g/HL mannoproteins inhibited bitartrate salt precipitation in wines even after having been kept at −4°C for 6 days. Yeast MP are efficient inhibitors at concentrations of 20 g/HL. However, for highly saturated wines, in which a higher concentration is needed to achieve the same inhibitory effect, MP flocculation may occur that counteracts the expected effect [93]. In a recent study, Guise et al. [94] reported that MP did not tartaric stabilized the wines. In fact, MP showed a variable effect, and thus needed preliminary tests to evaluate their effectiveness and the optimal dose, which was specific to the wine being treated [90, 95]. In conclusion, the effect of MP on tartaric acid stabilization is still a continuing matter of debate [94].

Wine MP can also modify wine aroma composition, either affecting the volatility and perception of wine aroma compounds or by releasing exogenous volatiles [96, 97]. The physicochemical interactions between aroma substances and MP depend on the nature of the volatile compounds, since a greater degree of interactions is often observed with hydrophobic compounds [96], as well as the conformational structure of the MP [12]. This fact implies a longer aromatic perception because the volatile compounds retained by MP will be slowly released [98, 99]. Some authors attribute the retention of the aroma substances to MP containing a high proportion of proteins as the protein fraction of MP is the main responsible for the aromatic stability [96]. However, Chalier et al. [12] have shown that both the glycosidic and peptidic parts of the MP may interact with the aroma compounds. Different authors have reported the role of yeast derivatives as a source of MP on wine aroma [97, 98, 100–102]. Dosage appears to be fundamental since low amounts of MP increased the volatility of some esters, giving more flowery and fruity notes to the wine; while higher amounts increased fatty acid content, producing yeasty, herbaceous and cheese-like smells [97]. In still wines, the use of free yeast strains with higher concentrations of MP resulted in higher concentration of positive aroma compounds, such as terpenes and C13-norisoprenoids associated with the fresh, fruity, and floral notes [103]. On the other hand, the addition of commercial products rich in MP in sparkling wines resulted in higher content of some fruity esters [102], and improved the perception of fruity [100, 101] and flowery characters [100]. It has been proposed that MP can be used to remove or reduce the occurrence of wine off-flavors as ethyl phenols (4-ethylguaiacol and 4-ethylphenol). In fact, the sorption of these compounds to the yeast walls could be due to the interactions of 4-ethylphenol and 4-ethylguaiacol with the functional groups of the MP and the free amino acids on the surface of the cell walls [104].

More interestingly, yeast MP have been described for their positive effect on the color stabilization [105, 107], reduction of astringency [10, 61, 108–110], and increased body and mouthfeel [10, 69, 99, 108, 111]. Studies performed in synthetic wines indicated that yeast MP can interact with tannins, probably through steric interactions, and prevent their aggregation and precipitation [7, 59]. This phenomenon seems to be dependent on the MP concentration and molecular weight, and on the conditions of the medium (ethanol content and ionic strength). The formation of tannin and polysaccharide complexes influences their association with salivary proteins, which then leads to a decrease in the astringency perception. This fact has been demonstrated in model solutions by several authors using different polysaccharide fractions [7, 10, 59]. It has also been evidenced not only the existence of interactions between MP and flavonols but also between MP and salivary proteins. This interaction could form proteins/polyphenol/mannoprotein soluble aggregates that probably affect the astringency perception [112]. Other studies suggested that MP did not stabilize or prevent the aggregation of tannin particles but they could increase tannin aggregation, leading to their precipitation [69, 108, 113]. The combination tannin-mannoproteins could result in high-molecular-weight structures that would be unstable and precipitate, leading to a decrease in the total proanthocyanidin content and thus, in a decrease in the astringent sensation [69, 108, 113]. More recently, Gonzalez Royo et al. [114] have shown that the decrease of the astringency sensation in wines was related to two different phenomena. The first was associated to the release of MP by inactive yeasts, which would increase the mouthfeel and inhibit the interactions between salivary proteins and tannins. The second was attributed to a direct effect of MP on the precipitation or absorption of proanthocyanidins. In fact, MP could act as stabilizers or flocculating polymers depending on factors such as tannin concentration and structure, and MP concentration, origin, molecular weight, charge, and structure [69]. It has also been reported that the addition of commercial inactive yeasts in grape juice during winemaking decreased the proanthocyanidin content of red wines coinciding with a decrease in high molecular weight MP [111, 115]. This fact suggests that the co-aggregates mannoprotein-tannin precipitated during this treatment [114]. Del Barrio-Galán et al. [99, 111] observed in the sensory analysis that some of MP commercial products reduced green tannins, thereby increasing softness on the palate. MP play also an important role in the stabilization of the color of red wines. MP are adsorbed by the colloidal molecules of anthocyanin-tannin, copigmented anthocyanins, and so forth, completely covering the surface of these colloids, avoiding their degradation and precipitation [116], leading to an increase in color stability [57, 105]. However, studies that analyze the effect of MP on wine color have shown contradictory results [7, 69, 99, 108, 109, 111, 113]. Our research group carried out a detailed study in order to know the effect of MP on the color of red wines. Several researches were carried out, such as the addition of commercial MP preparations before alcoholic fermentation [113], the use of MP overproducing yeast strains [69], aging on lysated lees [55], and combinations of all these treatments [55, 108]. Contrary to what was described in model solutions by using MP purified preparations [7, 59], our results showed that the use of MP in real vinification situations did not maintain the extracted polyphenols in colloidal dispersion, and neither seemed to ensure color stability [108, 113].

Other interesting oenological property of MP is their capacity to stimulate the growth of lactic acid bacteria and consequently the malolactic fermentation [117, 118]. In fact, MP can stimulate the malolactic bacteria through two mechanisms. Firstly, the adsorption of the medium chain fatty acids synthesized by Saccharomyces. These compounds have been shown to inhibit lactic acid bacteria growth and hence their removal by MP promotes the detoxification of the medium [119]. Secondly, the enzymatic hydrolysis of yeast MP and/or other macromolecules and polysaccharides by lactic acid bacteria can enhance the nutritional content of the medium, and thus potentially stimulate the lactic acid bacteria growth [117].

In the same way, yeast MP are able to adsorb the ochratoxin A (OTA), which is a dangerous mycotoxin [120]. This adsorption seems to be more effective in white wines than in red wines, due to the competition between polyphenols and OTA for the same binding sites on the surface of the yeast cells [106, 121]. There are several factors that can significantly affect the ability of OTA adsorption by MP as yeast strain [122, 123], mannosylphosphate content in the MP of wine yeasts, dissimilar fermentation, and cell sedimentation dynamics, cell dimension, and flocculence [120].

MP also affect the foam quality of sparkling wines [64, 65, 124–126]. Specifically, these molecules play a major role in foam stabilization [65, 127], particularly the MP with low content of protein (5%) [127]. The hydrophobic nature of MP causes them to preferentially adsorb to the gas/liquid interface of foam bubbles [128, 129], resulting in more stable foam [125]. In fact, the use of MP or cell wall extracts as additives has been proposed to improve the foam properties of sparkling wines elaborated by the traditional method. Therefore, the addition of yeast cell wall MP with a relative molecular weight between 10 and 30 kDa improved the foaming of sparkling wines [124]. However, the addition of commercial dry yeast products rich in MP to the tirage liquor did not modify the foam properties of sparkling wines [101]. In a previous work it was shown that MP and PRAG were poor foam formers but good foam stabilizers. Moreover, a higher positive correlation was found between foam stability time and PRAG (r = 0.723) than MP (r = 0.465) [65].

Finally, MP also contribute to the flocculation of yeast strains [130], and thus improve their elimination from the bottle during disgorging. MP could also serve as markers to follow the autolysis process because they are the major polysaccharides released by yeast [1, 3, 54]. Moreover, MP also seem to participate in film forming yeast or flor velum in Sherry type wines [131]. These wines are produced by "biological aging" that follows alcoholic fermentation. According to the study conducted by Alexandre et al. [132], a 49 kDa hydrophobic cell wall MP present in a velum yeast has been correlated with velum formation during the aging system used in sherry wine (Spain) or Vin Jaune (France).

4. Conclusions

Polysaccharides are one of the main groups of macromolecules in wines. They play an important role in both the technological and organoleptic properties of the wines. The oenological interest of polysaccharides has induced the development of several commercial products. In fact, there are nowadays in the market different commercial products based on purified MP or yeast derived cell walls, which are used in many wineries in order to improve the tartaric or proteic stability of the wines, or the sensory properties of some wines. However, these products have not always shown a clear effect in the wines. Recent studies indicate that other oligosaccharides and polysaccharide families from grapes could have a great potential to modify and improve the sensory and physicochemical properties of the wines. Unfortunately, these polysaccharide families are very difficult to obtain and they are not present in commercial formulates. Therefore, there are only a few studies regarding their effects and mechanisms of action, and more researches have to be done to better known their role and applicability into the wines.

Acknowledgements

The authors would like to thank the *Instituto Nacional de Investigación y Tecnología Agraria y Alimentaria (INIA)* for the funding provided for this study through the project RTA2017-00005-C02-02.

Author details

Leticia Martínez-Lapuente, Zenaida Guadalupe and Belén Ayestarán*
Institute of Vine and Wine Sciences, University of La Rioja, Government of La Rioja, CSIC, Logroño, Spain

*Address all correspondence to: belen.ayestaran@unirioja.es

References

[1] Doco T, Vuchot P, Cheynier V, Moutounet M. Structural modification of arabinogalactan-proteins during aging of red wines on lees. American Journal of Enology and Viticulture. 2003;**54**:150-157

[2] Guadalupe Z, Ayestarán B. Polysaccharide profile and content during the vinification and aging of tempranillo red wines. Journal of Agricultural and Food Chemistry. 2007;**55**:10720-10728. DOI: 10.1021/jf0716782

[3] Martínez-Lapuente L, Guadalupe Z, Ayestarán B, Ortega-Heras M, Pérez-Magariño S. Changes in polysaccharide composition during sparkling wine making and aging. Journal of Agricultural and Food Chemistry. 2013;**61**:12362-12373. DOI: 10.1021/jf403059p

[4] Apolinar-Valiente R, Romero-Cascales I, Williams P, Gómez Plaza E, López Roca JM, Ros-García JM, et al. Effect of winemaking techniques on polysaccharide composition of cabernet sauvignon, Syrah and Monastrell red wines. Australian Journal of Grape and Wine Research. 2014;**20**:62-71. DOI: 10.1111/ajgw.12048

[5] Martínez-Lapuente L, Guadalupe Z, Ayestarán B. Effect of egg albumin fining, progressive clarification and cross-flow microfiltration on the polysaccharide and proanthocyanidin composition of red varietal wines. Food Research International. 2017;**96**:235-243. DOI: 10.1016/j.foodres.2017.03.022

[6] Guadalupe Z. Manoproteínas y enzimas en la extracción y estabilidad del color de vinos tintos de tempranillo [thesis]. Logroño: Universidad de La Rioja; 2008

[7] Riou V, Vernhet A, Doco T, Moutounet M. Aggregation of grape seed tannins in model wine—Effect of wine polysaccharides. Food Hydrocolloids. 2002;**16**:17-23. DOI: 10.1016/S0268-005X(01)00034-0

[8] Waters E, Pellerin P, Brillouet JM. A saccharomyces mannoprotein that protects wine from protein haze. Carbohydrate Polymers. 1994;**23**:185-191. DOI: 10.1016/0144-8617(94)90101-5

[9] Lankhorst PP, Voogt B, Tuinier R, Lefol B, Pellerine P, Virone C. Prevention of tartrate crystallization in wine by hydrocolloids: The mechanism studied by dynamic light scattering. Journal of Agricultural and Food Chemistry. 2017;**65**:8923-8929. DOI: 10.1021/acs.jafc.7b01854

[10] Vidal S, Francis L, Williams P, Kwiatkowski M, Gawel R, Cheynier V, et al. The mouth-feel properties of polysaccharides and anthocyanins in a wine like medium. Food Chemistry. 2004;**85**:519-525. DOI: 10.1016/S0308-8146(03)00084-0

[11] Brandão E, Santos Silva M, García-Estévez I, Williams P, Mateus N, Doco T, et al. The role of wine polysaccharides on salivary protein-tannin interaction: A molecular approach. Carbohydrate Polymers. 2017;**177**:77-85. DOI: 10.1016/j.carbpol.2017.08.075

[12] Chalier P, Angot B, Delteil D, Doco T, Gunata Z. Interactions between aroma compounds and whole mannoprotein isolated from *Saccharomyces cerevisiae* strains. Food Chemistry. 2007;**100**:22-30. DOI: 10.1016/j.foodchem.2005.09.004

[13] Vidal S, Williams P, Doco T, Moutounet M, Pellerin P. The polysaccharides of red wine: Total fractionation and characterisation. Carbohydrate Polymers.

2003;**54**:439-447. DOI: 10.1016/S0144-8617(03)00152-8

[14] Vorwerk S, Somerville S, Somerville C. The role of plant cell polysaccharide composition in disease resistance. Trends in Plant Science. 2004;**9**:203-209. DOI: 10.1016/j.tplants.2004.02.005

[15] de Vries RP, Visser J. Aspergillus enzymes involved in degradation of plant cell wall polysaccharides. Microbiology and Molecular Biology Reviews. 2001;**65**:497-522. DOI: 10.1128/MMBR.65.4.497-522.2001

[16] Caffall KH, Mohnen D. The structure, function, and biosynthesis of plant cell wall pectic polysaccharides. Carbohydrate Research. 2009;**344**:1879-1900. DOI: 10.1016/j.carres.2009.05.021

[17] Guadalupe Z, Martínez-Pinilla O, Garrido A, Carrillo JD, Ayestarán B. Quantitative determination of wine polysaccharides by gas chromatography-mass spectrometry (GC-MS) and size exclusion chromatography (SEC). Food Chemistry. 2012;**131**:367-374. DOI: 10.1016/j.foodchem.2011.08.049

[18] Taiz L, Zeiger E. Plant Physiology. 3rd ed. Sunderland: Sinauer Associates; 2002. p. 313

[19] García E, Chacón JL, Martínez J, Izquierdo PM. Changes in volatile compounds during ripening in grapes of Airén, Macabeo and Chardonnay white varieties grown in La Mancha region (Spain). Food Science and Technology International. 2003;**9**:33-41. DOI: 10.1177/1082013203009001006

[20] Bindon KA, Madani SH, Pendleton P, Smith PA, Kennedy JA. Factors affecting skin tannin extractability in ripening grapes. Journal of Agricultural and Food Chemistry. 2014;**62**:1130-1141. DOI: 10.1021/jf4050606

[21] Pinelo M, Arnous A, Meyer AS. Upgrading of grape skins: Significance of plant cell-wall structural components and extraction techniques for phenol release. Trends in Food Science and Technology. 2006;**17**:579-590. DOI: 10.1016/j.tifs.2006.05.003

[22] Lecas M, Brillouet JM. Cell wall composition of grape berry skins. Phytochemistry. 1994;**35**:1241-1243. DOI: 10.1016/S0031-9422(00)94828-3

[23] Rid M, Markheiser A, Hoffmann C, Gross J. Waxy bloom on grape berry surface is one important factor for oviposition of European grapevine moths. Journal of Pest Science. 2018;**91**:1225-1239. DOI: 10.1007/s10340-018-0988-7

[24] Gao Y, Fangel JU, Willats WGT, Vivier MA, Moore JP. Dissecting the polysaccharide-rich grape cell wall changes during winemaking using combined high-throughput and fractionation methods. Carbohydrate Polymers. 2015;**133**:567-577. DOI: 10.1016/j.carbpol.2015.07.026

[25] Moore JP, Fangel JU, Willats WGT, Vivier MA. Pectic-β(1,4)-galactan, extensin and arabinogalactan-protein epitopes differentiate ripening stages in wine and table grape cell walls. Annals of Botany. 2014;**114**:1279-1294. DOI: 10.1093/aob/mcu053

[26] Ralet MC, Tranquet O, Poulain D, Moïse A, Guillon F. Monoclonal antibodies to rhamnogalacturonan I backbone. Planta. 2010;**231**:1373-1383. DOI: 10.1007/s00425-010-1116-y

[27] Doco T, Williams P, Pauly M, O'Neill MA, Pellerin P. Polysaccharides from grape berry cell walls part II. Structure characterization of the xyloglucan polysaccharides. Carbohydrate Polymers. 2003;**53**:253-261. DOI: 10.1016/S0144-8617(03)00072-9

[28] Coetzee ZA, Walker RR, Deloire A, Clarke SJ, Barril C, Rogiers SY. Spatiotemporal changes in the accumulation of sugar and potassium within individual'sauvignon Blanc (*Vitis vinifera* L.) berries'. Vitis—Journal of Grapevine Research. 2017;**56**:189-195. DOI: 10.5073/vitis.2017.56.193-195

[29] Ortega-Regules A, Romero-Cascales I, Ros-García JM, López-Roca JM, Gómez-Plaza E. A first approach towards the relationship between grape skin cell-wall composition and anthocyanin extractability. Analytica Chimica Acta. 2006;**563**:26-32. DOI: 10.1016/j.aca.2005.12.024

[30] Ortega-Regules A, Ros-García JM, Bautista-Ortín AB, López-Roca JM, Gómez-Plaza E. Differences in morphology and composition of skin and pulp cell walls from grapes (*Vitis vinifera* L.) technological implications. European Food Research and Technology. 2008;**227**:223-231. DOI: 10.1007/s00217-007-0714-9

[31] Fasoli M, Dell'Anna R, Dal Santo S, Balestrini R, Sanson A, Pezzotti M, et al. Pectins, hemicelluloses and celluloses show specific dynamics in the internal and external surfaces of grape berry skin during ripening. Plant & Cell Physiology. 2016;**57**:1332-1349. DOI: 10.1093/pcp/pcw080

[32] Fry SC. Primary cell wall metabolism: Tracking the careers of wall polymers in living plant cells. The New Phytologist. 2004;**161**:641-675. DOI: 10.1111/j.1469-8137.2004.00980.x

[33] Minic Z, Jouanin L. Plant glycoside hydrolases involved in cell wall polysaccharide degradation. Plant Physiology and Biochemistry. 2006;**44**:435-449. DOI: 10.1016/j.plaphy.2006.08.001

[34] Franková L, Fry SC. Biochemistry and physiological roles of enzymes that 'cut and paste' plant cell-wall polysaccharides. Journal of Experimental Botany. 2013;**64**:3519-3550. DOI: 10.1093/jxb/ert201

[35] Nunan KJ, Sims IM, Bacic A, Robinson SP, Fincher GB. Changes in cell wall composition during ripening of grape berries. Plant Physiology. 1998;**118**:783-792. DOI: 10.1104/pp.118.3.783

[36] Garrido-Bañuelos G, Buica A, Schückel J, Zietsman AJJ, Willats WGT, Moore JP, et al. Investigating the relationship between grape cell wall polysaccharide composition and the extractability of phenolic compounds into shiraz wines. Part I: Vintage and ripeness effects. Food Chemistry. 2019;**278**:36-46. DOI: 10.1016/j.foodchem.2018.10.134

[37] Bindon KA, Bacic A, Kennedy JA. Tissue-specific and developmental modifications of grape cell walls influence the adsorption of proanthocyanidins. Journal of Agricultural and Food Chemistry. 2012;**60**:9249-9260. DOI: 10.1021/jf301552t

[38] Silacci MW, Morrison JC. Changes in pectin content of cabernet sauvignon grape berries during maturation. American Journal of Enology and Viticulture. 1990;**41**:111-115

[39] Bordiga M, Travaglia F, Meyrand M, Bruce German J, Lebrilla CB, Coïsson JD, et al. Identification and characterization of complex bioactive oligosaccharides in white and red wine by a combination of mass spectrometry and gas chromatography. Journal of Agricultural and Food Chemistry. 2012;**60**:3700-3707. DOI: 10.1021/jf204885s

[40] Martínez-Lapuente L, Apolinar-Valiente R, Guadalupe Z, Ayestarán B, Pérez-Magariño S, Williams P, et al. Influence of grape maturity on complex carbohydrate

composition of red sparkling wines. Journal of Agricultural and Food Chemistry. 2016;**64**:5020-5030. DOI: 10.1021/acs.jafc.6b00207

[41] Guadalupe Z, Ayestarán B, Williams P, Doco T. Determination of must and wine polysaccharides by gas chromatography- mass spectrometry (GC-MS) and size-exclusion chromatography (SEC). In: Ramawat K, Mérillon JM, editors. Polysaccharides. Germany: Springer; 2014. pp. 1-28. DOI: 10.1007/978-3-319-03751-6_56-2

[42] Jégou S, Hoang DA, Salmon T, Williams P, Oluwa S, Vrigneau C, et al. Effect of grape juice press fractioning on polysaccharide and oligosaccharide compositions of pinot meunier and chardonnay champagne base wines. Food Chemistry. 2017;**232**:49-59. DOI: 10.1016/j.foodchem.2017.03.032

[43] Zamora F. La maceración prefermentativa en frío de la uva tinta. Enólogos. 2004;**32**:36-39

[44] Doco T, Williams P, Cheynier V. Effect of flash release and pectinolytic enzyme treatments on wine polysaccharide composition. Journal of Agricultural and Food Chemistry. 2007;**55**:6643-6649. DOI: 10.1021/jf071427t

[45] Cheynier V, Sarni-Manchado P. Wine taste and mouthfeel. In: Reynolds AG, editor. Managing Wine Quality. United Kingdom: Woodhead Publishing; 2010. pp. 29-72. DOI: 10.1533/9781845699284.1.30

[46] Boulton R. The copigmentation of anthocyanins and its role in the color of red wine: A critical review. American Journal of Enology and Viticulture. 2001;**52**:67-87

[47] Sánchez Palomo E, Díaz-Maroto Hidalgo MC, González-Viñas MA, Pérez-Coello MS. Aroma enhancement in wines from different grape varieties using exogenous glycosidases. Food Chemistry. 2005;**92**:627-635. DOI: 10.1016/j.foodchem.2004.08.025

[48] Ducasse MA, Canal-Llauberes RM, de Lumley M, Williams P, Souquet JM, Fulcrand H, et al. Effect of macerating enzyme treatment on the polyphenol and polysaccharide composition of red wines. Food Chemistry. 2010;**118**:369-376. DOI: 10.1016/j.foodchem.2009.04.130

[49] Apolinar-Valiente R, Romero-Cascales I, Williams P, Gómez- Plaza E, López-Roca JM, Ros-García JM, et al. Oligosaccharides of cabernet sauvignon, Syrah and Monastrell red wines. Food Chemistry. 2015;**179**:311-317. DOI: 10.1016/j.foodchem.2015.01.139

[50] Ayestarán B, Guadalupe Z, León D. Quantification of major grape polysaccharides (Tempranillo v.) released by maceration enzymes during the fermentation process. Analytica Chimica Acta. 2004;**513**:29-39. DOI: 10.1016/j.aca.2003.12.012

[51] Zimman A, Joslin W, Lyon M, Meier J, Waterhouse A. Maceration variables affecting phenolic composition in commercial-scale cabernet sauvignon winemaking trials. American Journal of Enology and Viticulture. 2002;**53**:93-98

[52] Álvarez I, Aleixandre JL, García MJ, Lizama V. Impact of prefermentative maceration on the phenolic and volatile compounds in Monastrell red wines. Analytica Chimica Acta. 2005;**563**:109-115. DOI: 10.1016/j.aca.2005.10.068

[53] Doco T, Brillouet JM. Isolation and characterisation of a rhamnogalacturonan II from red wine. Carbohydrate Research. 1993;**243**:333-343. DOI: 10.1016/0008-6215(93)87037-S

[54] Charpentier C, Dos Santos AM, Feuillat M. Release of macromolecules by *Saccharomyces cerevisiae* during

ageing of French flor sherry wine "vin jaune". International Journal of Food Microbiology. 2004;**96**:253-262. DOI: 10.1016/j.ijfoodmicro.2004.03.019

[55] Fernández O, Martínez O, Hernández Z, Guadalupe Z, Ayestarán B. Effect of the presence of lysated lees on polysaccharides, color and main phenolic compounds of red wine during barrel ageing. Food Research International. 2011;**44**:84-91. DOI: 10.1016/j.foodres.2010.11.008

[56] Pati S, Esti M, Leoni A, Liberatore MT, La Notte E. Polysaccharide and volatile composition of cabernet wine affected by different over-lees ageing. European Food Research and Technology. 2012;**235**:537-543. DOI: 10.1007/s00217-012-1781-0

[57] Ribéreau-Gayon P, Glories Y, Maujean A, Dubourdieu D. Handbook of Enology: The Chemistry of Wine, Stabilization and Treatments. Vol. 2. West Sussex: Wiley; 2006. DOI: 10.1002/0470010398

[58] Carvalho E, Mateus N, Plet B, Pianet I, Dufourc E, de Freitas V. Isolation and structural characterization of new anthocyanin-derived yellow pigments in aged red wines. Journal of Agricultural and Food Chemistry. 2006;**54**:8936-8944. DOI: 10.1021/jf062325q

[59] Poncet-Legrand C, Doco T, Williams P, Vernhet A. Inhibition of grape seed tannin aggregation by wine mannoproteins: Effect of polysaccharide molecular weight. American Journal of Enology and Viticulture. 2007;**58**:87-91

[60] Watrelot AA, Schulz DL, Kennedy JA. Wine polysaccharides influence tannin-protein interactions. Food Hydrocolloids. 2017;**63**:571-579. DOI: 10.1016/j.foodhyd.2016.10.010

[61] Quijada-Morín N, Williams P, Rivas-Gonzalo JC, Doco T, Escribano-Bailón MT. Polyphenolic, polysaccharide and oligosaccharide composition of Tempranillo red wines and their relationship with the perceived astringency. Food Chemistry. 2014;**154**:44-51. DOI: 10.1016/j.foodchem.2013.12.101

[62] Pellerin P, Waters EJ, Brillouet JM, Moutounet M. Effet de polysaccharides sur la formation de trouble protéique dans un vin blanc. Journal International des Sciences de la Vigne et du Vin. 1994;**28**:213-225. DOI: 10.20870/oeno-one.1994.28.3.1144

[63] Gerbaud V, Gabas N, Laguerie C, Blouin J, Vidal S, Moutounet M, et al. Effect of wine polysaccharides on the nucleation of potassium hydrogen tartrate in model solutions. Chemical Engineering Research and Design. 1996;**74**:782-789

[64] Vincenzi S, Crapisi A, Curioni A. Foamability of Prosecco wine: Cooperative effects of high molecular weight glycocompounds and wine PR proteins. Food Hydrocolloids. 2014;**34**:202-207. DOI: 10.1016/j.foodhyd.2012.09.016

[65] Martínez-Lapuente L, Guadalupe Z, Ayestarán B, Pérez-Magariño S. Role of major wine constituents in the foam properties of white and rosé sparkling wines. Food Chemistry. 2015;**174**:330-338. DOI: 10.1016/j.foodchem.2014.10.080

[66] Jouquand C, Ducruet V, Giampaoli P. Partition coefficients of aroma compounds in polysaccharide solutions by the phase ratio variation method. Food Chemistry. 2004;**85**:467-474. DOI: 10.1016/j.foodchem.2003.07.023

[67] Dufour C, Bayonove CL. Influence of wine structurally different polysaccharides on the volatility of aroma substances in a model system. Journal of Agricultural and Food

Chemistry. 1999;**47**:671-678. DOI: 10.1021/jf9801062

[68] Mitropoulou A, Hatzidimitriou E, Paraskevopoulou A. Aroma release of a model wine solution as influenced by the presence of non-volatile components. Effect of commercial tannin extracts, polysaccharides and artificial saliva. Food Research International. 2011;**44**:1561-1570. DOI: 10.1016/j.foodres.2011.04.023

[69] Guadalupe Z, Martínez L, Ayestarán B. Yeast mannoproteins in red winemaking: Effect on polysaccharide, polyphenolic, and color composition. American Journal of Enology and Viticulture. 2010;**61**:191-200

[70] Rosi I, Gheri A, Domizio P, Fia G. Production de macromolécules pariétales de *Saccharomyces cerevisiae* au cours de la fermentation et leur influence sur la fermentation malolactique. Revue des Oenologues et des Techniques Vitivinicoles et Oenologiques. 2000;**94**:18-20

[71] Pérez-Serradilla JA, Luque de Castro MD. Role of lees in wine production: A review. Food Chemistry. 2008;**111**:447-456. DOI: 10.1016/j.foodchem.2008.04.019

[72] Klis FM, Boorsma A, De Groot PWJ. Cell wall construction in *Saccharomyces cerevisiae*. Yeast. 2006;**23**:185-202. DOI: 10.1002/yea.1349

[73] Saulnier LT, Mercereau T, Vezinhet F. Mannoproteins from flocculating and non-flocculating *Saccharomyces cerevisiae* yeast. Journal of the Science of Food and Agriculture. 1991;**54**:275-286. DOI: 10.1002/jsfa.2740540214

[74] Vernhet A, Pellerin P, Prieur C, Osmianski J, Moutounet M. Charge properties of some grape and wine polysaccharide and polyphenolic fractions. American Journal of Enology and Viticulture. 1996;**47**:25-29

[75] Samant SK, Singhal RS, Kulkaml PR, Rege DV. Protein-polysaccharide interactions: A new approach in food formulation. International Journal of Food Science and Technology. 1993;**28**:547-562. DOI: 10.1111/j.1365-2621.1993.tb01306.x

[76] Caridi A. Enological functions of parietal yeast mannoproteins. Antonie Van Leeuwenhoek. 2006;**89**:417-422. DOI: 10.1007/s10482-005-9050-x

[77] Vernhet A, Pellerin P, Belleville MP, Planque J, Moutounet M. Relative impact of major wine polysaccharides on the performances of an organic microfiltration membrane. American Journal of Enology and Viticulture. 1999;**50**:51-56

[78] Morata A, Gomez-Cordoves MC, Suberviola J, Bartolome B, Colomo B, Suarez JA. Adsorption of anthocyanins by yeast cell walls during the fermentation of red wines. Journal of Agricultural and Food Chemistry. 2003;**51**:4084-4088. DOI: 10.1021/jf021134u

[79] Ferreira RB, Picarra-Pereira MA, Monteiro S, Loureiro VB, Teixeira AR. The wine proteins. Trends in Food Science and Technology. 2002;**12**:230-239. DOI: 10.1016/S0924-2244(01)00080-2

[80] Pocock KF, Waters EJ. Protein haze in bottled white wines: How well do stability tests and bentonite fining trials predict haze formation during storage and transport? Australian Journal of Grape and Wine Research. 2006;**12**:212-220. DOI: 10.1111/j.1755-0238.2006.tb00061.x

[81] Esteruelas M, Poinsaut P, Sieczkowski N, Manteau S, Fort MF, Canals JM, et al. Characterization of natural haze protein in sauvignon white wine. Food Chemistry. 2009;**113**:28-35. DOI: 10.1016/j.foodchem.2008.07.031

[82] Moine-Ledoux V, Dubourdieu D. An invertase fragment responsible for improving the protein stability of dry white wines. Journal of the Science of Food and Agriculture. 1999;**79**:537-543. DOI: 10.1002/(SICI)1097-0010(19990315)79:4<537::AID-JSFA214>3.0.CO;2-B

[83] Dupin IV, McKinnon BM, Ryan C, Boulay M, Markides AJ, Jones GP, et al. *Saccharomyces cerevisiae* mannoproteins that protect wine from protein haze: Their release during fermentation and lees contact and a proposal for their mechanism of action. Journal of Agricultural and Food Chemistry. 2000;**48**:3098-3105. DOI: 10.1021/jf0002443

[84] Waters EJ, Wallace W, Tate ME, Williams PJ. Isolation and partial characterization of a natural haze protective factor from wine. Journal of Agricultural and Food Chemistry. 1993;**41**:724-730. DOI: 10.1021/jf00029a009

[85] Gazzola D, Van Sluyter SC, Curioni A, Waters EJ, Marangon M. Roles of proteins, polysaccharides, and phenolics in haze formation in white wine via reconstitution experiments. Journal of Agricultural and Food Chemistry. 2012;**60**:10666-–10673. DOI: 10.1021/jf302916n

[86] Lomolino G, Curioni A. Protein haze formation in white wines: Effect of *Saccharomyces cerevisiae* cell wall components prepared with different procedures. Journal of Agricultural and Food Chemistry. 2007;**55**:8737-8744. DOI: 10.1021/jf0712550

[87] Dufrechou M, Doco T, Poncet-Legrand C, Sauvage FX, Vernhet A. Protein/polysaccharide interactions and their impact on haze formation in white wines. Journal of Agricultural and Food Chemistry. 2015;**63**:10042-10053. DOI: 10.1021/acs.jafc.5b02546

[88] Schmidt SA, Tan EL, Brown S, Nasution UJ, Pettolino F, Macintyre OJ, et al. Hpf2 glycan structure is critical for protection against protein haze formation in white wine. Journal of Agricultural and Food Chemistry. 2009;**57**:3308-3315. DOI: 10.1021/jf803254s

[89] Ribeiro T, Fernandes C, Nunes FM, Filipe-Ribeiro L, Cosme F. Influence of the structural features of commercial mannoproteins in white wine protein stabilization and chemical and sensory properties. Food Chemistry. 2014;**159**:47-54. DOI: 10.1016/j.foodchem.2014.02.149

[90] Moine-Ledoux V, Dubourdieu D. Role of yeast mannoproteins with regard to tartaric stabilisation of wines. Bulletin de l'Organisation Internationale de la Vigne et du Vin. 2002;**75**:471-483

[91] Gerbaud V, Gabas N, Blouin J, Pellerin P, Moutounet M. Influence of wine polysaccharides and polyphenols on the crystallization of potassium hydrogen tartrate. Journal International des Sciences de la Vigne et du Vin. 1997;**31**:65-83. DOI: 10.20870/oeno-one.1997.31.2.1087

[92] Moutounet M, Battle JL, Saint Pierre B, Escudier JL. Stabilisation tartrique. Détermination du degré d'instabilité des vins. Mesure de l'efficacité des inhibiteurs de cristallisation. In: Lonvaud-Funel A, editor. Oenologie 1999. 6° Symposium International d'Oenologie. Paris: Lavoisier Tec-Doc; 1999. pp. 531-534

[93] Gerbaud V, Gabas N, Blouin J, Crachereau JC. Study of wine tartaric acid salt stabilization by addition of carboxymethylcellulose (CMC): Comparison with the "protective colloids" effect. Journal International des Sciences de la Vigne et du Vin. 2010;**44**:231-242. DOI: 10.20870/oeno-one.2010.44.4.1474

[94] Guise R, Filipe-Ribeiro L, Nascimento D, Bessa O, Nunes FM, Cosme F. Comparison between different

types of carboxylmethylcellulose and other oenological additives used for white wine tartaric stabilization. Food Chemistry. 2014;**156**:250-257. DOI: 10.1016/j.foodchem.2014.01.081

[95] Bosso A, Panero L, Petrozziello M, Sollazzo M, Asproudi A, Motta S, et al. Use of polyaspartate as inhibitor of tartaric precipitations in wines. Food Chemistry. 2015;**185**:1-6. DOI: 10.1016/j.foodchem.2015.03.099

[96] Lubbers S, Voilley A, Feuillat M, Charpentier C. Influence of mannoproteins from yeast on the aroma intensity of a model wine. LWT-Food Science and Technology. 1994;**27**:108-114. DOI: 10.1006/fstl.1994.1025

[97] Comuzzo P, Tat L, Tonizzo A, Battistutta F. Yeast derivatives (extracts and autolysates) in winemaking: Release of volatile compounds and effects on wine aroma volatility. Food Chemistry. 2006;**99**:217-230. DOI: 10.1016/j.foodchem.2005.06.049

[98] Del Barrio-Galán R, Pérez-Magariño S, Ortega-Heras M, Williams P, Doco T. Effect of aging on lees and of three different dry yeast derivative products on Verdejo white wine composition and sensorial characteristics. Journal of Agricultural and Food Chemistry. 2011;**59**:12433-12442. DOI: 10.1021/jf204055u

[99] Del Barrio-Galán R, Pérez-Magariño S, Ortega-Heras M, Guadalupe Z, Ayestarán B. Polysaccharide characterization of commercial dry yeast preparations and their effect on white and red wine composition. LWT-Food Science and Technology. 2012;**48**:215-223. DOI: 10.1016/j.lwt.2012.03.016

[100] Rodriguez-Nogales JM, Fernández-Fernández E, Vila-Crespo J. Effect of the addition of b-glucanases and commercial yeast preparations on the chemical and sensorial characteristics of traditional sparkling wine. European Food Research and Technology. 2012;**235**:729-744. DOI: 10.1007/s00217-012-1801-0

[101] Pérez-Magariño S, Martínez-Lapuente L, Bueno-Herrera M, Ortega-Heras M, Guadalupe Z, Ayestarán B. Use of commercial dry yeast products rich in mannoproteins for white and rosé sparkling wine elaboration. Journal of Agricultural and Food Chemistry. 2015;**63**:5670-5681. DOI: 10.1021/acs.jafc.5b01336

[102] Costa GP, Nicolli KP, Welke JE, Manfroi V, Zini CA. Volatile profile of sparkling wines produced with the addition of mannoproteins or lees before second fermentation performed with free and immobilized yeasts. Journal of the Brazilian Chemical Society. 2018;**1866-1875**:29. DOI: 10.21577/0103-5053.20180062

[103] Juega M, Carrascosa AV, Martinez-Rodriguez AJ. Effect of short ageing on lees on the mannoprotein content, aromatic profile, and sensorial character of white wines. Journal of Food Science. 2015;**80**:384-388. DOI: 10.1111/1750-3841.12763

[104] Nieto-Rojo R, Ancín-Azpilicueta C, Garrido JJ. Sorption of 4-ethylguaiacol and 4-ethylphenol on yeast cell walls, using a synthetic wine. Food Chemistry. 2014;**152**:399-406. DOI: 10.1016/j.foodchem.2013.11.157

[105] Escot S, Feuillat M, Dulau L, Charpentier C. Release of polysaccharides by yeast and the influence of released polysaccharides on color stability and wine astringency. Australian Journal of Grape and Wine Research. 2001;7:153-159. DOI: 10.1111/j.1755-0238.2001.tb00204.x

[106] Feuillat M, Escot S, Charpentier C. Élevage des vins rouges sur lies fines-Intérêt des interactions entre polysaccharides de levure et polyphénols du vin. Revue des Oenologues et

des Techniques Vitivinicoles et Oenologicques: Magazine Trimestriel d'information Professionnelle. 2001;**98**:17-18

[107] Gonçalves FJ, Wessel DF, Cardoso SM, Rocha SM, Coimbra MA. Interaction of wine mannoproteins and arabinogalactans with anthocyanins. Food Chemistry. 2018;**243**:1-10. DOI: 10.1016/j.foodchem.2017.09.097

[108] Guadalupe Z, Palacios A, Ayestarán B. Maceration enzymes and mannoproteins: A possible strategy to increase colloidal stability and color extraction in red wines. Journal of Agricultural and Food Chemistry. 2007;**55**:4854-4862. DOI: 10.1021/jf063585a

[109] Rodrigues A, Ricardo-Da-Silva JM, Laureano O. Effect of commercial mannoproteins on wine colour and tannins stability. Food Chemistry. 2012;**131**:907-914. DOI: 10.1016/j.foodchem.2011.09.075

[110] Diako C, McMahon K, Mattinson S, Evans M, Ross C. Alcohol, tannins, and mannoprotein and their interactions influence the sensory properties of selected commercial merlot wines: A preliminary study. Journal of Food Science. 2016;**81**:2039-2048. DOI: 10.1111/1750-3841.13389

[111] Del Barrio-Galán R, Pérez-Magariño S, Ortega-Heras M. Techniques for improving or replacing ageing on lees of oak aged red wines: The effects on polysaccharides and the phenolic composition. Food Chemistry. 2011;**127**:528-540. DOI: 10.1016/j.foodchem.2011.01.035

[112] Ramos-Pineda AM, García-Estévez I, Dueñas M, Escribano-Bailón MT. Effect of the addition of mannoproteins on the interaction between wine flavonols and salivary proteins. Food Chemistry. 2018;**264**:226-232. DOI: 10.1016/j.foodchem.2018.04.119

[113] Guadalupe Z, Ayestarán B. Effect of commercial mannoproteins addition on polysaccharide, polyphenolic and color composition in red wines. Journal of Agricultural and Food Chemistry. 2008;**9022-9029**:56. DOI: 10.1021/jf801535k

[114] González-Royo E, Esteruelas M, Kontoudakis N, Fort F, Canals JM, Zamora F. The effect of supplementation with three commercial inactive dry yeasts on the colour, phenolic compounds, polysaccharides and astringency of a model wine solution and red wine. Journal of the Science of Food and Agriculture. 2017;**97**:172-181. DOI: 10.1002/jsfa.7706

[115] Fornairon-Bonnefond C, Camarasa C, Moutounet M, Salmón JM. New trends on yeast autolysis and wine ageing on lees: A bibliographical review. Journal International des Sciences de la Vigne et du Vin. 2002;**36**:49-69. DOI: 10.20870/oeno-one.2002.36.2.974

[116] Feuillat M, Charpentier C. Les mannoproteins de levure. Un adjuvant oenologique possible. Bulletin de l'Organisation Internationale de la Vigne et du Vin. 1998;**71**:929-940

[117] Guilloux-Benatier M, Chassagne D. Comparison of components released by fermented or active dried yeasts after aging on lees in a model wine. Journal of Agricultural and Food Chemistry. 2003;**51**:746-751. DOI: 10.1021/jf020135j

[118] Díez L, Guadalupe Z, Ayestarán B, Ruiz-Larrea F. Effect of yeast mannoproteins and grape polysaccharides on the growth of wine lactic acid and acetic acid bacteria. Journal of Agricultural and Food Chemistry. 2010;**58**:7731-7739. DOI: 10.1021/jf100199n

[119] Alexandre H, Costello PJ, Remize F, Guzzo J, Guilloux-Benatier M. *Saccharomyces cerevisiae*-Oenococcus oeni interactions in wine: Current

knowledge and perspectives. International Journal of Food Microbiology. 2004;**93**:141-154. DOI: 10.1016/j.ijfoodmicro.2003.10.013

[120] Caridi A. New perspectives in safety and quality enhancement of wine through selection of yeasts based on the parietal adsorption activity. International Journal of Food Microbiology. 2007;**120**:167-172. DOI: 10.1016/j.ijfoodmicro.2007.08.032

[121] Ummarino I, García-Moruno E, Di Estefano R. Interazione polifenoliscorze di lievito. Rivista di Viticoltura e di Enologia. 2001;**4**:37-46

[122] Cecchini F, Morassut M, Garcia Moruno E, Di Stefano R. Influence of yeast strain on ochratoxin a content during fermentation of white and red must. Food Microbiology. 2006;**23**:411-417. DOI: 10.1016/j.fm.2005.08.003

[123] Caridi A, Galvano F, Tafuri A, Ritieni A. Ochratoxin a removal during winemaking. Enzyme and Microbial Technology. 2006;**40**:122-126. DOI: 10.1016/j.enzmictec.2006.07.002

[124] Núñez YP, Carrascosa AV, González R, Polo MC, Martínez-Rodríguez A. Isolation and characterization of a thermally extracted yeast cell wall fraction potentially useful for improving the foaming properties of sparkling wines. Journal of Agricultural and Food Chemistry. 2006;**54**:7898-7903. DOI: 10.1021/jf0615496

[125] Blasco L, Viñas M, Villa TG. Proteins influencing foam formation in wine and beer: The role of yeast. International Microbiology. 2011;**14**: 61-71. DOI: 10.2436/20.1501.01.136

[126] Martínez-Lapuente L, Ayestarán B, Guadalupe Z. Influence of wine chemical compounds on the foaming properties of sparkling wines. In: Jordão AM, Cosme F, editors. Grapes and Wines. Advances in Production Processing Analysis and Valorization. Rijeka, Croatia: Intechopen; 2017. pp. 195-223. DOI: 10.5772/intechopen.70859

[127] Coelho E, Rocha SM, Coimbra MA. Foamability and foam stability of molecular reconstituted model sparkling wines. Journal of Agricultural and Food Chemistry. 2011;**59**:8770-8778. DOI: 10.1021/jf2010657

[128] Moreno-Arribas V, Pueyo E, Nieto FJ, Martín-Álvarez PJ, Polo MC. Influence of the polysaccharides and nitrogen compounds on foaming properties of sparkling wines. Food Chemistry. 2000;**70**:309-317. DOI: 10.1016/S0308-8146(00)00088-1

[129] Martínez-Rodríguez A, Carrascosa AV, Martín-Álvarez PJ, Moreno- Arribas V, Polo MC. Influence of the yeast strain on the changes of the amino acids, peptides and proteins during sparkling wine production by the traditional method. Journal of Industrial Microbiology and Biotechnology. 2002;**29**:314-322. DOI: 10.1038/sj.jim.7000323

[130] Klis FM, Mol P, Hellingwerf K, Brul S. Dynamics of cell wall structure in *Saccharomyces cerevisiae*. FEMS Microbiology Reviews. 2002;**26**:239-256. DOI: 10.1111/j.1574-6976.2002.tb00613.x

[131] Legras JL, Moreno-García J, Zara S, Zara G, Garcia-Martinez T, Mauricio JC, et al. Flor yeast: New perspectives beyond wine aging. Frontiers in Microbiology. 2016;**7**:1-11. DOI: 10.3389/fmicb .2016.00503

[132] Alexandre H, Blanchet S, Charpentier C. Identification of a 49-kDa hydrophobic cell wall mannoprotein present in velum yeast which may be implicated in velum formation. FEMS Microbiology Letters. 2000;**185**:147-150. DOI: 10.1111/j.1574-6968.2000.tb09053.x

Chapter 7

Flavonoids and Pectins

Zhiping Zhang, Yanzhi He and Xinyue Zhang

Abstract

Pectins and flavonoids are two related groups of important secondary metabolites derived from plants. The interaction between pectins and flavonoids can affect their shelf-life stability, functionality, bioavailability, and bioaccessibility. In this chapter, we will concentrate on the current opinions on the flavonoids to understand how to classify this group of secondary metabolites, what biological and pharmacological activities they possess, and how to biosynthesize them in plants. We will then discuss the general strategies for the derivation of these small secondary compounds. The strategies comprise traditional plant extraction, chemical synthesis, and biosynthesis. We will also discuss the advantages and disadvantages of these three production strategies in the derivation of flavonoids and the future research directions in generating health-beneficial flavonoids using the biosynthetic strategy.

Keywords: flavonoids, pectins, secondary metabolites, interaction, pharmacological activity, biological activity, biosynthetic pathway, extraction, characterization, chemical synthesis, microbial cell factory, enzymatic synthesis, multienzyme synthetic system

1. Introduction

Peels represent a large percentage of the total weight of fruits, for example, 50–65% of *Citrus* fruits (lemon, lime, orange, and grapefruit) [1]. During processing of fruits for juice and oil extractions, the peels remain as the primary byproducts and become waste if not processed further, which can lead to serious environmental pollution [1]. Therefore, the fruit-processing industries are also interested in making use of these wastes.

The peels are also a good commercial source of pectins (polygalacturonic acid) and flavonoids [1]. The pectins are polysaccharide macromolecules contained in the primary cell wall of plants and involved in controlling cell wall ionic status, cell expansion, and separation [1]. Usually, the pectins are commercially extracted and isolated from *Citrus* peels and apple pomace. They are not only used as a gelling agent, dessert filling, or juice and milk stabilizer in food industry but also as a source of dietary fiber. Flavonoids are a large group of small secondary metabolites contained in the vacuoles and possess a wide range of biological activities, especially those with human health benefits [2, 3]. In the *Citrus* peels, flavonoids mainly include flavones (e.g., rhoifolin, isorhoifolin, diosmin, and neodiosmin), flavanones (e.g., eriocitrin, neoeriocitrin, narirutin, naringin, hesperidin, neohesperidin, poncirin, and neoponcirin), and flavonols (e.g., rutin) [4]. It has been reported that the highest concentrations of *Citrus* flavonoids occur in the peels [1]. Due to the importance of pectins and flavonoids in food, cosmetic, and medicinal industries, quite a number of studies have been focused on these two groups of compounds. Accordingly, a variety of approaches have been developed for efficient isolation of pectins and flavonoids from fruit peels and pomace. For example, to make a better use of yellow passion fruit rind, de Souza and colleagues have developed a strategy for sequential extraction of flavonoids and pectin [5].

As we know, pectins are abundant in the middle lamella of the plant cell walls with a gradual decrease in the content toward the plasma membrane, whereas flavonoids are naturally located within the cells [6]. Generally, flavonoids within the cells do not come into contact with the cell wall materials, such as pectins, celluloses, and hemicelluloses, prior to food processing. When fruits are processed and eaten, intracellular flavonoids can be released from the cells, leading to their interaction with substances like metal ions and plant cell wall materials [7, 8]. For example, procyanidins and anthocyanins can spontaneously bind to water-, chelator-, and sodium carbonate-soluble pectins. It is believed that the binding of flavonoids to cell wall materials results from noncovalent, hydrophobic, hydrogen bonding, and ionic interactions [9–11]. Recently, Chirug and colleagues have presented a novel possible mechanism that iron ions mediate the interaction between pectins and quercetin [6]. Such interaction might affect their shelf-life stability and functionality, as well as their bioavailability and bioaccessibility [6, 12]. Therefore, it could be of high importance to study their interaction. Since there are several reviews on the interaction [8, 13], we will not discuss it in this chapter. Instead, we will concentrate on understanding the current opinions on flavonoids, including the classification, biological activities, and biosynthetic pathway of these secondary compounds. We will then review the general strategies for derivation of these compounds, including the traditional plant extraction, chemical synthesis, and biosynthesis of these important small bioactive molecules in a microbial cell factory or an *in vitro* multienzyme synthetic platform. We will also discuss the advantages and disadvantages of these strategies and the future research directions in the field of flavonoid biosynthesis.

2. Classification and biological activities of flavonoids

Flavonoids belong to a class of secondary metabolites and comprise a large group of natural products that are widespread in higher plants but also found in mosses and liverworts [14, 15]. Chemically, flavonoid compounds have the basic structure of 15-carbon atoms with two phenolic rings connected by a 3-carbon chain [16], forming a C6-C3-C6 carbon framework (**Figure 1**). Generally, these small molecules can be divided into six major subclasses on the basis of the variations on the

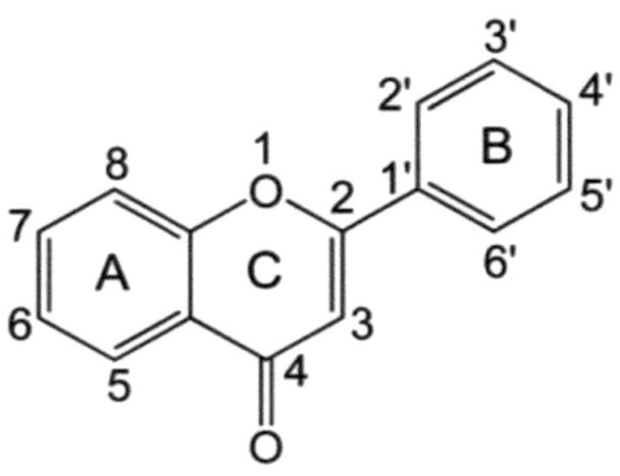

Figure 1.
Structure and atom numbering of flavonoid backbone.

heterocyclic C-ring and the degree of oxidation: the flavanols, flavones, flavonols, flavanones, anthocyanidins, and isoflavones [2, 16–18]. The flavonoids can exist in a free aglycone form but are often glycosylated (most commonly glucose), and the glycosylation in turn increases their water solubility [19].

The flavonoids are involved in the formation of plant pigments [20] and protect plants against pathogens, herbivores, and UV radiation [21]. However, the study of flavonoids, like that of most natural products, has emerged from the search of new compounds with promising pharmacological properties. After decades of endeavors, scientists have found that flavonoids possess a wide variety of biological and pharmacological properties, which leads to numerous studies on these secondary metabolites. These health-beneficial properties include antiangiogenic [22], antibacterial [23–27], anti-cancer [24, 28], anti-inflammatory [28–33], antiglycating [34], antimalarial [35], antimicrobial [36–42], anti-oxidant [26, 36, 38, 42–51], anti-platelet [48], anti-proliferation [52], agonistic/antagonistic [53], ammonia-lowering and regulation of urea cycle [54], anxiolytic [55], atheroprotective [56], cardioprotective and hypouricemic [57], cytotoxic [51, 58], endocrine disrupting [59], free radical-scavenging [31–33, 39, 40, 46, 51, 52, 58, 60–66], hepatoprotective [67], leishmanicidal [68], neuroprotective [69], photoprotective [43], and trypanocidal activities [68, 70]. In addition, the flavonoids can inhibit eukaryotic protein synthesis [71] and a variety of important enzymes such as aggrecanase [72], aldose reductase [30, 73], alpha-glucosidase [60], cholinesterase [26, 74], protein tyrosine phosphatase and acetylcholinesterase [75], and tyrosinase [44, 64].

3. Biosynthetic pathway of flavonoids

After several decades of efforts, the pathway for flavonoid biosynthesis has been largely deciphered even though quite a number of details remain unknown (**Figure 2**). The flavonoids and their derivatives are biosynthesized by a variety of enzymes. These enzymes belong to different families [76], mainly including 2-oxoglutarate-dependent dioxygenase (2-ODD), cytochrome P450 hydroxylase, short-chain dehydrogenase/reductase (SDR), *O*-methyltransferase (OMT), and *O*-glycosyltransferase (GT). The 2-ODD, cytochrome P450, and SDR enzymes constitute the major pathway for flavonoid biosynthesis [76], and the OMT and GT enzymes are involved in modification of flavonoids. The involved 2-ODD enzymes mainly comprise flavanone 3-hydroxylase (F3H), flavonol synthase (FLS), flavone synthase I (FSI), anthocyanidin synthase (ANS), and flavonol 6-hydroxylase (F6H) [17, 76–81]. The related cytochrome P450 enzymes contain cinnamate 4-hydroxylase (C4H), isoflavone synthase (IFS), flavanone 2-hydroxylase (F2H), flavone synthase II (FSII), flavonol 6-hydroxylase (F6H), flavonoid 3′-hydroxylase (F3’H), flavonoid 3′,5′-hydroxylase (F3′5′H), isoflavone 2′-hydroxylase (I2′H), and isoflavone 3′-hydroxylase (I3′H) [17, 18, 76, 80, 82, 83]. The SDR enzymes participating in flavonoid biosynthesis include dihydroflavonol 4-reductase (DFR) and anthocyanidin synthase (ANR) [76]. Interestingly, the flavone synthase (FS) activity is specified either by a 2-ODD (FSI) or a P450 (FSII) enzyme in a plant species-dependent manner [84, 85]. Similarly, the flavonol 6-hydroxylase (F6H) activity is also endowed either by a 2-ODD [81, 86] or P450 [87, 88] enzyme in different plant species. These findings further increase the complexity of flavonoid biosynthesis.

Basically, biosynthesis of flavonoids can be arbitrarily divided into three major stages. The first stage (a.k.a the phenylpropanoid pathway) includes three successive chemical reactions catalyzed by phenylalanine ammonia-lyase (PAL), cinnamate 4-hydroxylase (C4H), and 4-coumaroyl:CoA ligase (4CL), respectively, to convert L-phenylalanine to 4-coumroyl-CoA. In addition, L-tyrosine can also

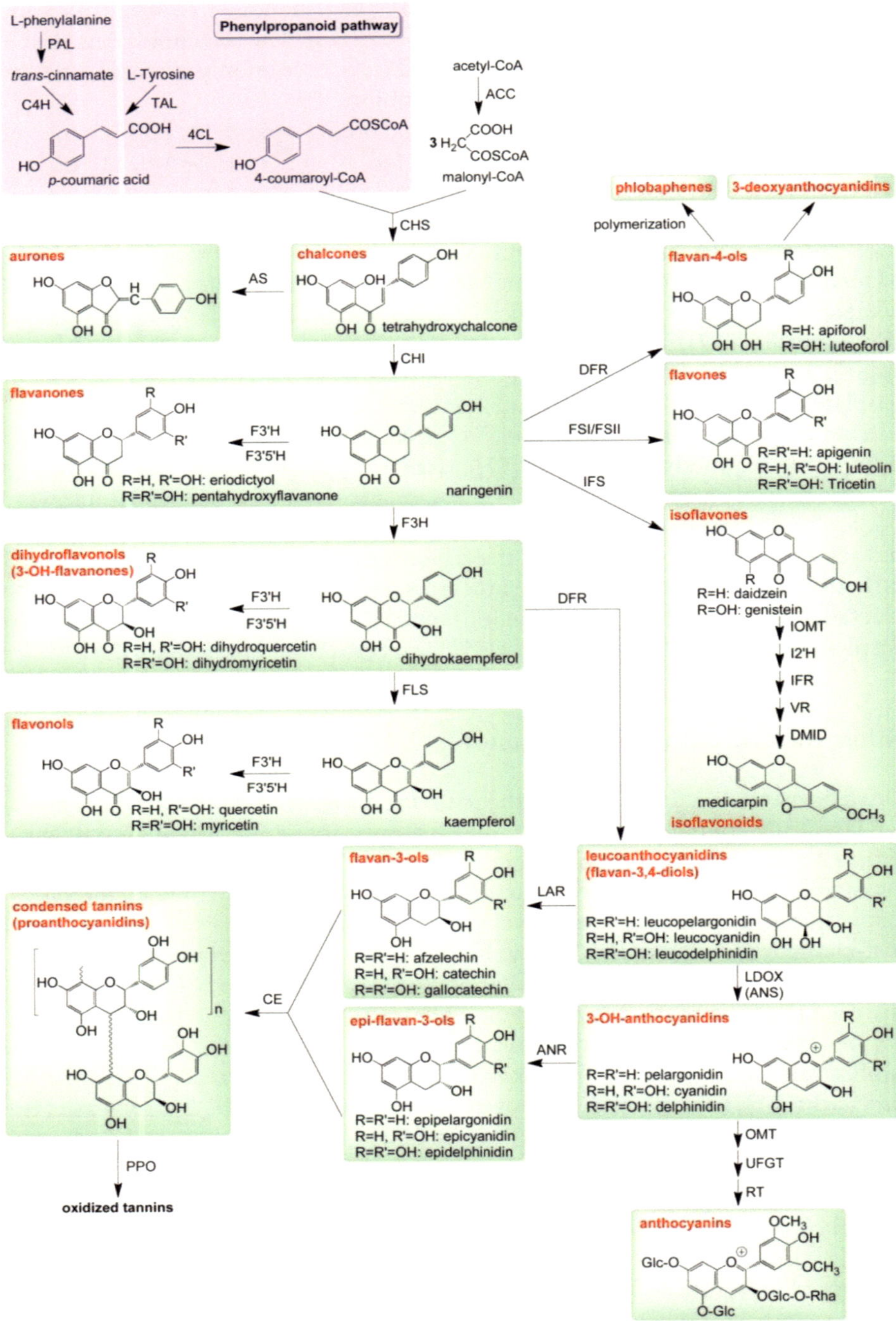

Figure 2.

Schematic of the biosynthetic pathway leading to the major subclasses of flavonoids. Adapted from [10, 12, 68]. 4CL, 4-coumaroyl:CoA ligase; ACC, acetyl CoA carboxylase; ANR, anthocyanidin reductase; ANS, anthocyanidin synthase; AS, aureusidin synthase; C4H, cinnamate 4-hydroxylase; CE: condensing enzyme; CHI, chalcone isomerase; CHS, chalcone synthase; DFR, dihydroflavonol 4-reductase; DMID, 7,2'-dihydroxy-4'-methoxyisoflavanol dehydratase; F3H, flavanone 3-hydroxylase; F3'H, flavonoid 3'-hydroxylase; F3'5'H, flavonoid 3',5'-hydroxylase; FLS, flavonol synthase; FSI/FSII, flavone synthase I/II; I2'H, isoflavone 2'-hydroxylase; IFR, isoflavone reductase; IFS, isoflavone synthase; IOMT, isoflavone O-methyltransferase; LAR, leucoanthocyanidin reductase; LDOX, leucoanthocyanidin dioxygenase; OMT, O-methyltransferase; PAL, phenylalanine ammonia-lyase; PPO, polyphenol oxidase; RT, rhamnosyltransferase; TAL, tyrosine ammonia-lyase; UFGT, UDP flavonoid glucosyltransferase; VR, vestitone reductase.

participate in the flavonoid biosynthesis via two successive enzymatic reactions catalyzed by tyrosine ammonia lyase (TAL) and 4CL, respectively. The second stage is crucial for the biosynthesis of flavonoids, in which the backbones of major subclasses of flavonoids are formed. This stage begins from the formation of chalcone by conversion of the 4-coumroyl-CoA from the first stage and the malonyl-CoA from carboxylation of acetyl-CoA. Chalcone synthase (CHS), an entry point enzyme into the pathway, catalyzes this chemical reaction by conversion of one molecule of 4-coumroyl-CoA and three molecules of malonyl-CoA to one molecule of chalcone (e.g., tetrahydroxychalcone). Then, the chalcone molecule is cyclized to form a flavanone (e.g., naringenin) by chalcone isomerase (CHI) and an aurone (e.g., aureusidin) by aureusidin synthase (AS). The flavanone can be further converted to dihydroflavonol by F3H and then flavonol by FLS. Alternatively, the flavanone molecule can also be converted to a flavone by FS, a flavanol by DFR, an isoflavone by IFS, and an anthocyanidin by a series of successive enzymatic reactions catalyzed by F3H, DFR, and leucoanthocyanidin dioxygenase (LDOX), respectively. The resulting anthocyanidin molecule can be further modified to form anthocyanins by a series of chemical modifications by OMT, UDP flavonoid glucosyltransferase (UFGT), and rhamnosyltransferase (RT). The third stage is mainly involved in various chemical decorations of flavonoids. Generally, natural flavonoids are often extensively modified by chemical reactions, including glycosylation and methylation [76], acylation [89], sulfonation [90, 91], prenylation [92, 93], and galloylation [94], which further contribute to the structural and functional diversity of flavonoids.

4. Derivation of flavonoids

Due to the intrinsic health benefits possessed by flavonoids, numerous approaches have been developed during the past decades for the derivation of a wide range of flavonoids. Basically, these approaches can be divided into three major categories: traditional plant extraction, chemical synthesis, and biosynthesis.

4.1 Traditional plant extraction via organic solvents

Traditionally, flavonoids are extracted from various plant species, which currently remains the most commonly used methods. During the past decades, researchers have developed plenty of methods to improve the yield and purity of flavonoids derived from plants. Generally, the plant tissues are air-dried and ground into powder for extraction via organic solvents (most commonly methanol and ethanol), and the extracts are then subjected to successive fractionation with other organic solvents (most commonly petroleum ether, chloroform, ethyl acetate, and n-butyl alcohol), followed by repeated silica gel and Sephadex LH-20 column chromatographies [44, 95]. The yield of plant-derived flavonoids can be improved by ultrasonic wave- [96], microwave- [97], and enzyme-assisted extraction [98]; aqueous two-phase extraction [99]; and a combination of these modifications [100]. The isolated flavonoids are then subjected to polyamide thin plate chromatography (TLC), high performance liquid chromatography (HPLC), electrospray ionization mass spectrometry (ESI-MS), and nuclear magnetic resonance (NMR) analyses to determine their identity and purity [2, 3]. Due to the high solubility of most flavonoids in organic solvents, this strategy often demonstrates a high efficiency in the derivation of flavonoids from plant tissues. However, the disadvantage of the plant extraction is obvious. Due to the very low content of most flavonoids in plant tissues, the extraction and isolation of flavonoids often requires multiple steps and plenty of time, labor, and organic solvents, which greatly increase the production cost. Moreover, different plant tissues often need to develop different approaches for processing, which makes the extraction more complicated and further increase the cost for the production of flavonoids. Therefore, this approach is not cost-effective, and it is crucial to develop alternative strategies to reduce the cost for producing flavonoids.

4.2 Chemical synthesis of flavonoids

Another approach for producing flavonoids is chemical synthesis. Basically, there are two strategies for chemical synthesis of flavones, that is, the chalcone route and the Baker-Venkataraman method [101]. Even though there are a few successful examples, chemical synthesis of flavonoids is often very complicated and involved in many steps [2]. It requires toxic reagents and extreme reaction conditions [3, 102]. Chiral synthesis and subsequent modifications further increase the difficulty of this approach in the production of flavonoids [3]. Moreover, the multistep chemical reactions often produce quite a number of intermediate products with a high similarity in structure, which further increases the difficulty in purification of the desired products. Therefore, chemical synthesis is not economically feasible for the mass production of flavonoids [3].

4.3 Biosynthesis of flavonoids

Since the biosynthetic pathway of flavonoids is largely elucidated in plants [20], other promising alternative strategies have been developed to produce these secondary compounds [2, 103–106]. One of these alternative strategies is to produce flavonoids in a microbial cell factory. It has been well known that *Escherichia coli* and *Saccharomyces cerevisiae* are the two most commonly used model organisms for the construction of a microbial cell factory. There are quite a few paradigms for the production of flavonoids using this strategy. For example, eriodictyol has been produced using L-tyrosine as a substrate in *E. coli* BL21(DE3) genetically modified by *TAL, 4CL, CHS, CHI, F3H*, and *F3'H* genes and the production can reach up to 107 mg/L by further introducing three other genes *acs*, *accBC*, and *dtsR1* to enhance the availability of malonyl-CoA [103]. Kaempferol has been produced in a microbial cell factory by introducing a *de novo* biosynthetic pathway into *S. cerevisiae*, and the biosynthesis has been further improved by introducing two more pathways to enhance the generation of acetyl-CoA and malonyl-CoA [107]. Obviously, this strategy circumvents some inherent disadvantages of traditional plant extraction and chemical synthesis. However, not all genetically modified microbes can produce desired products due to the well-known complexity of a microbial cell system, the incompatibility of artificially synthesized genetic elements in host cells, the growth inhibition of host cells by desired and intermediate products, and the instability of an engineered biosystem itself [2, 108].

Recently, we have developed an *in vitro* platform to produce flavonoids by constructing a multienzyme synthetic system to convert naringenin into kaempferol in one pot [2]. After optimizing a series of reaction parameters, including the components and pH value of the buffer system, reaction temperature and time, and total amount and ratio of the enzymes, the production yield can reach up to 37.55 ± 1.62 mg/L within 40–50 min with a conversion rate of 55.89% ± 2.74% [2]. The advantages of this strategy are obvious. It is time- and labor-saving. The reaction conditions are easy to control accurately. Due to the clearness in the buffer components and the lack of complex physiological regulation as occurred in the microbial cell factory, it is possible to easily make further optimization in the future.

DOI: http://dx.doi.org/10.5772/intechopen.84960

It is also much easier to purify desired products from this *in vitro* synthetic system than from the cell factory because of the simplicity of the components in the system. In addition, the strategy is highly cost-effective because of the cheap chemicals and recombinant proteins used in this system. More importantly, the system is easy to scale up and therefore possesses a huge industrialization potential. It also provides a guide for other secondary metabolites to produce economically. However, problems still exist in this production strategy. For example, due to the lack of P450-reductase function, prokaryotically expressed cytochrome P450 enzymes lose their enzymatic activities [109]. To achieve a functional expression, Leonard and colleagues fused a plant P450 enzyme gene *F3′5′H* with its redox partner cytochrome P450 reductase gene *cpr* from *Catharanthus roseus* and successfully produced a hydroxylated flavonol quercetin from *p*-coumaric acid in *E. coli* by simultaneous coexpression of the fusion protein with 4CL, CHS, CHI, F3H, and FLS [110], which provides a guide to solve this kind of problem. To further improve the efficiency of the biosynthetic system, future research should be focused on screening key enzymes with high activities from various plants, mutation of genes encoding key enzymes to enhance their activities, and immobilization of the highly active enzymes to inert carriers.

5. Conclusions

Pectins and flavonoids are two distinctive classes of bioactive secondary metabolites presented in the fruit peels and used in food industry. The flavonoids can be divided into six major subclasses, including the flavanols, flavones, flavonols, flavanones, anthocyanidins, and isoflavones, and their flavonoid biosynthetic pathway has been largely elucidated. These natural small compounds possess a wide range of health-beneficial properties and can be derived by traditional plant extraction via organic solvents, chemical synthesis, and biosynthesis by constructing a microbial cell factory or an *in vitro* multienzyme synthetic system.

Conflict of interest

The authors declare that they have no competing financial interests.

Author details

Zhiping Zhang[1], Yanzhi He[1] and Xinyue Zhang[1,2,3,4]*

1 College of Bioscience and Biotechnology, Yangzhou University, Yangzhou City, Jiangsu, China

2 Key Laboratory of Prevention and Control of Biological Hazard Factors (Animal Origin) for Agrifood Safety and Quality, Ministry of Agriculture of China, Yangzhou University, Yangzhou City, Jiangsu, China

3 Joint International Research Laboratory of Agriculture & Agri-Product Safety, Yangzhou University, Yangzhou City, Jiangsu, China

4 Jiangsu Co-innovation Center for Prevention and Control of Important Animal Infectious Diseases and Zoonoses, Yangzhou City, Jiangsu, China

*Address all correspondence to: zhangxinyue@yzu.edu.cn

References

[1] Mandalari G, Bennett RN, Bisignano G, Saija A, Dugo G, Lo Curto RB, et al. Characterization of flavonoids and pectins from bergamot (*Citrus bergamia* Risso) peel, a major byproduct of essential oil extraction. Journal of Agricultural and Food Chemistry. 2006;**54**(1):197-203. DOI: 10.1021/jf051847n

[2] Zhang Z, He Y, Huang Y, Ding L, Chen L, Liu Y, et al. Development and optimization of an *in vitro* multienzyme synthetic system for production of kaempferol from naringenin. Journal of Agricultural and Food Chemistry. 2018;**66**(31):8272-8279. DOI: 10.1021/acs.jafc.8b01299

[3] Malla S, Pandey RP, Kim BG, Sohng JK. Regiospecific modifications of naringenin for astragalin production in *Escherichia coli*. Biotechnology and Bioengineering. 2013;**110**(9):2525-2535. DOI: 10.1002/bit.24919

[4] Nogata Y, Sakamoto K, Shiratsuchi H, Ishii T, Yano M, Ohta H. Flavonoid composition of fruit tissues of citrus species. Bioscience, Biotechnology, and Biochemistry. 2006;**70**(1):178-192. DOI: 10.1271/bbb.70.178

[5] de Souza CG, Rodrigues TH, LM ES, Ribeiro PR, de Brito ES. Sequential extraction of flavonoids and pectin from yellow passion fruit rind using pressurized solvent or ultrasound. Journal of the Science of Food and Agriculture. 2018;**98**(4):1362-1368. DOI: 10.1002/jsfa.8601

[6] Chirug L, Okun Z, Ramon O, Shpigelman A. Iron ions as mediators in pectin-flavonols interactions. Food Hydrocolloids. 2018;**84**:441-449. DOI: 10.1016/j.foodhyd.2018.06.039

[7] El Hajji H, Nkhili E, Tomao V, Dangles O. Interactions of quercetin with iron and copper ions: Complexation and autoxidation. Free Radical Research. 2006;**40**(3):303-320. DOI: 10.1080/10715760500484351

[8] Zhu F. Interactions between cell wall polysaccharides and polyphenols. Critical Reviews in Food Science and Nutrition. 2018;**58**(11):1808-1831. DOI: 10.1080/10408398.2017.1287659

[9] Lin Z, Fischer J, Wicker L. Intermolecular binding of blueberry pectin-rich fractions and anthocyanin. Food Chemistry. 2016;**194**:986-993. DOI: 10.1016/j.foodchem.2015.08.113

[10] Le Bourvellec C, Bouchet B, Renard CM. Non-covalent interaction between procyanidins and apple cell wall material. Part III: Study on model polysaccharides. Biochimica et Biophysica Acta. 2005;**1725**(1):10-18. DOI: 10.1016/j.bbagen.2005.06.004

[11] Le Bourvellec C, Guyot S, Renard C. Non-covalent interaction between procyanidins and apple cell wall material: Part I. Effect of some environmental parameters. Biochimica et Biophysica Acta. 2004;**1672**:192-202. DOI: 10.1016/j.bbagen.2004.04.001

[12] Phan AD, Netzel G, Wang D, Flanagan BM, D'Arcy BR, Gidley MJ. Binding of dietary polyphenols to cellulose: Structural and nutritional aspects. Food Chemistry. 2015;**171**:388-396. DOI: 10.1016/j.foodchem.2014.08.118

[13] Le Bourvellec C, Renard CMGC. Interactions between polyphenols and macromolecules: Quantification methods and mechanisms. Critical Reviews in Food Science and Nutrition. 2012;**52**(1-3):213-248. DOI: 10.1080/10408398.2010.499808

[14] Buckingham J, Munasinghe VRN. Dictionary of Flavonoids: With

CD-ROM. Boca Raton: CRC Press, Taylor & Francis Group; 2015. 1169 p

[15] Panche AN, Diwan AD, Chandra SR. Flavonoids: An overview. Journal of Nutritional Science. 2016;**5**:e47. DOI: 10.1017/jns.2016.41

[16] Mulvihill EE, Burke AC, Huff MW. Citrus flavonoids as regulators of lipoprotein metabolism and atherosclerosis. Annual Review of Nutrition. 2016;**36**:275-299. DOI: 10.1146/annurev-nutr-071715-050718

[17] Tohge T, de Souza LP, Fernie AR. Current understanding of the pathways of flavonoid biosynthesis in model and crop plants. Journal of Experimental Botany. 2017;**68**(15): 4013-4028. DOI: 10.1093/jxb/erx177

[18] Lepiniec L, Debeaujon I, Routaboul JM, Baudry A, Pourcel L, Nesi N, et al. Genetics and biochemistry of seed flavonoids. Annual Review of Plant Biology. 2006;**57**:405-430. DOI: 10.1146/annurev.arplant.57.032905.105252

[19] Croft KD. The chemistry and biological effects of flavonoids and phenolic acids. Annals of the New York Academy of Sciences. 1998;**854**:435-442. DOI: 10.1111/j.1749-6632.1998.tb09922.x

[20] Winkel-Shirley B. Flavonoid biosynthesis. A colorful model for genetics, biochemistry, cell biology, and biotechnology. Plant Physiology. 2001;**126**(2):485-493. DOI: 10.1104/pp.126.2.485

[21] Heim KE, Tagliaferro AR, Bobilya DJ. Flavonoid antioxidants: Chemistry, metabolism and structure-activity relationships. The Journal of Nutritional Biochemistry. 2002;**13**(10):572-584. DOI: 10.1016/S0955-2863(02)00208-5

[22] Hassan LE, Ahamed MB, Majid AS, Baharetha HM, Muslim NS, Nassar ZD, et al. Correlation of antiangiogenic, antioxidant and cytotoxic activities of some Sudanese medicinal plants with phenolic and flavonoid contents. BMC Complementary and Alternative Medicine. 2014;**14**:406. DOI: 10.1186/1472-6882-14-406

[23] Kurnia D, Apriyanti E, Soraya C, Satari MH. Antibacterial flavonoids against oral bacteria of *Enterococcus faecalis* ATCC 29212 from Sarang Semut (*Myrmecodia pendans*) and its inhibitor activity against enzyme MurA. Current Drug Discovery Technologies. 2019;**16**:1-7. DOI: 10.2174/1570163815666180828113920

[24] Albadawi DA, Mothana RA, Khaled JM, Ashour AE, Kumar A, Ahmad SF, et al. Antimicrobial, anticancer, and antioxidant compounds from *Premna resinosa* growing in Saudi Arabia. Pharmaceutical Biology. 2017;**55**(1):1759-1766. DOI: 10.1080/13880209.2017.1322617

[25] Chakotiya AS, Tanwar A, Narula A, Sharma RK. *Zingiber officinale*: Its antibacterial activity on *Pseudomonas aeruginosa* and mode of action evaluated by flow cytometry. Microbial Pathogenesis. 2017;**107**:254-260. DOI: 10.1016/j.micpath.2017.03.029

[26] Bensouici C, Kabouche A, Karioti A, Ozturk M, Duru ME, Bilia AR, et al. Compounds from *Sedum caeruleum* with antioxidant, anticholinesterase, and antibacterial activities. Pharmaceutical Biology. 2016;**54**(1):174-179. DOI: 10.3109/13880209.2015.1028078

[27] Biva IJ, Ndi CP, Griesser HJ, Semple SJ. Antibacterial constituents of Eremophila alternifolia: An Australian aboriginal traditional medicinal plant. Journal of Ethnopharmacology. 2016;**182**:1-9. DOI: 10.1016/j.jep.2016.02.011

[28] Zhou H, Lutterodt H, Cheng Z, Yu LL. Anti-inflammatory and antiproliferative activities of trifolirhizin, a flavonoid from *Sophora*

flavescens roots. Journal of Agricultural and Food Chemistry. 2009;**57**(11): 4580-4585. DOI: 10.1021/jf900340b

[29] Zhang L, Wu T, Xiao W, Wang Z, Ding G, Zhao L. Enrichment and purification of total ginkgo flavonoid O-glycosides from *Ginkgo biloba* extract with macroporous resin and evaluation of anti-inflammation activities *in vitro*. Molecules. 2018;**23**(5):1167. DOI: 10.3390/molecules23051167

[30] Han QT, Ren Y, Li GS, Xiang KL, Dai SJ. Flavonoid alkaloids from Scutellaria moniliorrhiza with anti-inflammatory activities and inhibitory activities against aldose reductase. Phytochemistry. 2018;**152**:91-96. DOI: 10.1016/j.phytochem.2018.05.001

[31] Chen XM, Tait AR, Kitts DD. Flavonoid composition of orange peel and its association with antioxidant and anti-inflammatory activities. Food Chemistry. 2017;**218**:15-21. DOI: 10.1016/j.foodchem.2016.09.016

[32] Diaz P, Jeong SC, Lee S, Khoo C, Koyyalamudi SR. Antioxidant and anti-inflammatory activities of selected medicinal plants and fungi containing phenolic and flavonoid compounds. Chinese Medicine. 2012;**7**(1):26. DOI: 10.1186/1749-8546-7-26

[33] Zhang L, Ravipati AS, Koyyalamudi SR, Jeong SC, Reddy N, Smith PT, et al. Antioxidant and anti-inflammatory activities of selected medicinal plants containing phenolic and flavonoid compounds. Journal of Agricultural and Food Chemistry. 2011;**59**(23): 12361-12367. DOI: 10.1021/jf203146e

[34] Ahmad R, Ahmad N, Naqvi AA, Exarchou V, Upadhyay A, Tuenter E, et al. Antioxidant and antiglycating constituents from leaves of *Ziziphus oxyphylla* and *Cedrela serrata*. Antioxidants (Basel). 2016;**5**(1):9. DOI: 10.3390/antiox5010009

[35] Chander MP, Pillai CR, Sunish IP, Vijayachari P. Antimicrobial and antimalarial properties of medicinal plants used by the indigenous tribes of Andaman and Nicobar Islands, India. Microbial Pathogenesis. 2016;**96**:85-88. DOI: 10.1016/j.micpath.2016.04.017

[36] Ng KR, Lyu X, Mark R, Chen WN. Antimicrobial and antioxidant activities of phenolic metabolites from flavonoid-producing yeast: Potential as natural food preservatives. Food Chemistry. 2019;**270**:123-129. DOI: 10.1016/j.foodchem.2018.07.077

[37] Tagousop CN, Tamokou JD, Ekom SE, Ngnokam D, Voutquenne-Nazabadioko L. Antimicrobial activities of flavonoid glycosides from Graptophyllum grandulosum and their mechanism of antibacterial action. BMC Complementary and Alternative Medicine. 2018;**18**(1):252. DOI: 10.1186/s12906-018-2321-7

[38] Pandey G, Khatoon S, Pandey MM, Rawat AKS. Altitudinal variation of berberine, total phenolics and flavonoid content in *Thalictrum foliolosum* and their correlation with antimicrobial and antioxidant activities. Journal of Ayurveda and Integrative Medicine. 2017;**9**(3):169-176. DOI: 10.1016/j.jaim.2017.02.010

[39] Silva-Beltran NP, Ruiz-Cruz S, Cira-Chavez LA, Estrada-Alvarado MI, Ornelas-Paz Jde J, Lopez-Mata MA, et al. Total phenolic, flavonoid, tomatine, and tomatidine contents and antioxidant and antimicrobial activities of extracts of tomato plant. International Journal of Analytical Chemistry. 2015;**2015**:284071. DOI: 10.1155/2015/284071

[40] Jalal TK, Ahmed IA, Mikail M, Momand L, Draman S, Isa ML, et al. Evaluation of antioxidant, total phenol and flavonoid content and antimicrobial activities of *Artocarpus altilis* (breadfruit) of underutilized tropical

fruit extracts. Applied Biochemistry and Biotechnology. 2015;**175**(7):3231-3243. DOI: 10.1007/s12010-015-1499-0

[41] Abdel Ghani SB, Weaver L, Zidan ZH, Ali HM, Keevil CW, Brown RC. Microwave-assisted synthesis and antimicrobial activities of flavonoid derivatives. Bioorganic & Medicinal Chemistry Letters. 2008;**18**(2):518-522. DOI: 10.1016/j.bmcl.2007.11.081

[42] Ayub MA, Hussain AI, Hanif MA, Chatha SAS, Kamal GM, Shahid M, et al. Variation in phenolic profile, beta-carotene and flavonoid contents, biological activities of two *Tagetes* species from Pakistani Flora. Chemistry & Biodiversity. 2017;**14**(6):e1600463. DOI: 10.1002/cbdv.201600463

[43] Lefahal M, Zaabat N, Ayad R, Makhloufi EH, Djarri L, Benahmed M, et al. In vitro assessment of total phenolic and flavonoid contents, antioxidant and photoprotective activities of crude methanolic extract of aerial parts of *Capnophyllum peregrinum* (L.) Lange (Apiaceae) growing in Algeria. Medicine. 2018;**5**(2):26. DOI: 10.3390/medicines5020026

[44] Kishore N, Twilley D, Blom van Staden A, Verma P, Singh B, Cardinali G, et al. Isolation of flavonoids and flavonoid glycosides from *Myrsine africana* and their inhibitory activities against mushroom tyrosinase. Journal of Natural Products. 2018;**81**(1):49-56. DOI: 10.1021/acs.jnatprod.7b00564

[45] Wang L, Wu Y, Bei Q, Shi K, Wu Z. Fingerprint profiles of flavonoid compounds from different *Psidium guajava* leaves and their antioxidant activities. Journal of Separation Science. 2017;**40**(19):3817-3829. DOI: 10.1002/jssc.201700477

[46] Badral D, Odonbayar B, Murata T, Munkhjargal T, Tuvshintulga B, Igarashi I, et al. Flavonoid and galloyl glycosides isolated from *Saxifraga spinulosa* and their antioxidative and inhibitory activities against species that cause piroplasmosis. Journal of Natural Products. 2017;**80**(9):2416-2423. DOI: 10.1021/acs.jnatprod.7b00142

[47] Abarikwu SO, Olufemi PD, Lawrence CJ, Wekere FC, Ochulor AC, Barikuma AM. Rutin, an antioxidant flavonoid, induces glutathione and glutathione peroxidase activities to protect against ethanol effects in cadmium-induced oxidative stress in the testis of adult rats. Andrologia. 2017;**49**(7):e12696. DOI: 10.1111/and.12696

[48] Assefa AD, Ko EY, Moon SH, Keum YS. Antioxidant and antiplatelet activities of flavonoid-rich fractions of three citrus fruits from Korea. 3 Biotech. 2016;**6**(1):109. DOI: 10.1007/s13205-016-0424-8

[49] Wang M, Xing S, Luu T, Fan M, Li X. The gastrointestinal tract metabolism and pharmacological activities of grosvenorine, a major and characteristic flavonoid in the fruits of *Siraitia grosvenorii*. Chemistry & Biodiversity. 2015;**12**(11):1652-1664. DOI: 10.1002/cbdv.201400397

[50] Vukics V, Kery A, Bonn GK, Guttman A. Major flavonoid components of heartsease (*Viola tricolor* L.) and their antioxidant activities. Analytical and Bioanalytical Chemistry. 2008;**390**(7):1917-1925. DOI: 10.1007/s00216-008-1885-3

[51] Mohammed RS, Abou Zeid AH, El Hawary SS, Sleem AA, Ashour WE. Flavonoid constituents, cytotoxic and antioxidant activities of *Gleditsia triacanthos* L. leaves. Saudi Journal of Biological Sciences. 2014;**21**(6):547-553. DOI: 10.1016/j.sjbs.2014.02.002

[52] Kim H, Choi HK, Moon JY, Kim YS, Mosaddik A, Cho SK. Comparative antioxidant and antiproliferative activities of red and white pitayas and

their correlation with flavonoid and polyphenol content. Journal of Food Science. 2011;**76**(1):C38-C45. DOI: 10.1111/j.1750-3841.2010.01908.x

[53] Van der Heiden E, Bechoux N, Muller M, Sergent T, Schneider YJ, Larondelle Y, et al. Food flavonoid aryl hydrocarbon receptor-mediated agonistic/antagonistic/synergic activities in human and rat reporter gene assays. Analytica Chimica Acta. 2009;**637**(1-2):337-345. DOI: 10.1016/j.aca.2008.09.054

[54] Nohara K, Shin Y, Park N, Jeong K, He B, Koike N, et al. Ammonia-lowering activities and carbamoyl phosphate synthetase 1 (Cps1) induction mechanism of a natural flavonoid. Nutrition & Metabolism (London). 2015;**12**:23. DOI: 10.1186/s12986-015-0020-7

[55] Li H, Zhou P, Yang Q, Shen Y, Deng J, Li L, et al. Comparative studies on anxiolytic activities and flavonoid compositions of *Passiflora edulis* 'edulis' and *Passiflora edulis* 'flavicarpa'. Journal of Ethnopharmacology. 2011;**133**(3):1085-1090. DOI: 10.1016/j.jep.2010.11.039

[56] Lara-Guzman OJ, Tabares-Guevara JH, Leon-Varela YM, Alvarez RM, Roldan M, Sierra JA, et al. Proatherogenic macrophage activities are targeted by the flavonoid quercetin. The Journal of Pharmacology and Experimental Therapeutics. 2012;**343**(2):296-306. DOI: 10.1124/jpet.112.196147

[57] Zhu JX, Wang Y, Kong LD, Yang C, Zhang X. Effects of *Biota orientalis* extract and its flavonoid constituents, quercetin and rutin on serum uric acid levels in oxonate-induced mice and xanthine dehydrogenase and xanthine oxidase activities in mouse liver. Journal of Ethnopharmacology. 2004;**93**(1):133-140. DOI: 10.1016/j.jep.2004.03.037

[58] El-Alfy TS, El-Gohary HM, Sokkar NM, Hosny M, Al-Mahdy DA. A new flavonoid C-glycoside from *Celtis australis* L. and *Celtis occidentalis* L. leaves and potential antioxidant and cytotoxic activities. Scientia Pharmaceutica. 2011;**79**(4):963-975. DOI: 10.3797/scipharm.1108-19

[59] Nordeen SK, Bona BJ, Jones DN, Lambert JR, Jackson TA. Endocrine disrupting activities of the flavonoid nutraceuticals luteolin and quercetin. Hormones and Cancer. 2013;**4**(5):293-300. DOI: 10.1007/s12672-013-0150-1

[60] Xiao X, Xu L, Hu H, Yang Y, Zhang X, Peng Y, et al. DPPH radical scavenging and postprandial hyperglycemia inhibition activities and flavonoid composition analysis of hawk tea by UPLC-DAD and UPLC-Q/TOF MS(E). Molecules. 2017;**22**(10):1622. DOI: 10.3390/molecules22101622

[61] Ismail HF, Hashim Z, Soon WT, Rahman NSA, Zainudin AN, Majid FAA. Comparative study of herbal plants on the phenolic and flavonoid content, antioxidant activities and toxicity on cells and zebrafish embryo. Journal of Traditional and Complementary Medicine. 2017;7(4):452-465. DOI: 10.1016/j.jtcme.2016.12.006

[62] Chakraborty S, Basu S. Multi-functional activities of citrus flavonoid narirutin in Alzheimer's disease therapeutics: An integrated screening approach and in vitro validation. International Journal of Biological Macromolecules. 2017;**103**:733-743. DOI: 10.1016/j.ijbiomac.2017.05.110

[63] Xi W, Fang B, Zhao Q, Jiao B, Zhou Z. Flavonoid composition and antioxidant activities of Chinese local pummelo (*Citrus grandis* Osbeck.) varieties. Food Chemistry. 2014;**161**:230-238. DOI: 10.1016/j.foodchem.2014.04.001

[64] Liang CP, Chang CH, Liang CC, Hung KY, Hsieh CW. In vitro antioxidant activities, free radical scavenging capacity, and tyrosinase inhibitory of flavonoid compounds and ferulic acid from *Spiranthes sinensis* (Pers.) Ames. Molecules. 2014;**19**(4):4681-4694. DOI: 10.3390/molecules19044681

[65] Li P, Jia J, Zhang D, Xie J, Xu X, Wei D. In vitro and in vivo antioxidant activities of a flavonoid isolated from celery (*Apium graveolens* L. var. dulce). Food & Function. 2014;**5**(1):50-56. DOI: 10.1039/c3fo60273g

[66] Ekuadzi E, Dickson R, Fleischer T, Annan K, Pistorius D, Oberer L, et al. Flavonoid glycosides from the stem bark of *Margaritaria discoidea* demonstrate antibacterial and free radical scavenging activities. Phytotherapy Research. 2014;**28**(5):784-787. DOI: 10.1002/ptr.5053

[67] Fernando CD, Soysa P. Total phenolic, flavonoid contents, in-vitro antioxidant activities and hepatoprotective effect of aqueous leaf extract of *Atalantia ceylanica*. BMC Complementary and Alternative Medicine. 2014;**14**:395. DOI: 10.1186/1472-6882-14-395

[68] Beer MF, Frank FM, German Elso O, Ernesto Bivona A, Cerny N, Giberti G, et al. Trypanocidal and leishmanicidal activities of flavonoids isolated from *Stevia satureiifolia* var. *satureiifolia*. Pharmaceutical Biology. 2016;**54**(10):2188-2195. DOI: 10.3109/13880209.2016.1150304

[69] Zhang YC, Gan FF, Shelar SB, Ng KY, Chew EH. Antioxidant and Nrf2 inducing activities of luteolin, a flavonoid constituent in Ixeris sonchifolia Hance, provide neuroprotective effects against ischemia-induced cellular injury. Food and Chemical Toxicology. 2013;**59**: 272-280. DOI: 10.1016/j.fct.2013.05.058

[70] Mergia E, Shibeshi W, Terefe G, Teklehaymanot T. Antitrypanosomal activity of *Verbascum sinaiticum* Benth. (Scrophulariaceae) against *Trypanosoma congolense* isolates. BMC Complementary and Alternative Medicine. 2016;**16**:362. DOI: 10.1186/s12906-016-1346-z

[71] Itoh A, Tanahashi T, Nagakura N, Takenaka Y, Chen CC, Pelletier J. Flavonoid glycosides from *Adina racemosa* and their inhibitory activities on eukaryotic protein synthesis. Journal of Natural Products. 2004;**67**(3): 427-431. DOI: 10.1021/np030440e

[72] Moncada-Pazos A, Obaya AJ, Viloria CG, Lopez-Otin C, Cal S. The nutraceutical flavonoid luteolin inhibits ADAMTS-4 and ADAMTS-5 aggrecanase activities. Journal of Molecular Medicine (Berlin, Germany). 2011;**89**(6):611-619. DOI: 10.1007/s00109-011-0741-7

[73] Mok SY, Lee S. Identification of flavonoids and flavonoid rhamnosides from *Rhododendron mucronulatum* for albiflorum and their inhibitory activities against aldose reductase. Food Chemistry. 2013;**136**(2):969-974. DOI: 10.1016/j.foodchem.2012.08.091

[74] Khan MT, Orhan I, Senol FS, Kartal M, Sener B, Dvorska M, et al. Cholinesterase inhibitory activities of some flavonoid derivatives and chosen xanthone and their molecular docking studies. Chemico-Biological Interactions. 2009;**181**(3):383-389. DOI: 10.1016/j.cbi.2009.06.024

[75] Nguyen DH, Seo UM, Zhao BT, Le DD, Seong SH, Choi JS, et al. Ellagitannin and flavonoid constituents from *Agrimonia pilosa* Ledeb. with their protein tyrosine phosphatase and acetylcholinesterase inhibitory activities. Bioorganic Chemistry. 2017;**72**:293-300. DOI: 10.1016/j.bioorg.2017.04.017

[76] Winkel BSJ. The biosynthesis of flavonoids. In: Grotewold E, editor. The

Science of Flavonoids. New York, NY: Springer New York; 2006. pp. 71-95

[77] Tohge T, Watanabe M, Hoefgen R, Fernie AR. The evolution of phenylpropanoid metabolism in the green lineage. Critical Reviews in Biochemistry and Molecular Biology. 2013;**48**(2):123-152. DOI: 10.3109/10409238.2012.758083

[78] Araujo WL, Tohge T, Nunes-Nesi A, Daloso DM, Nimick M, Krahnert I, et al. Phosphonate analogs of 2-oxoglutarate perturb metabolism and gene expression in illuminated Arabidopsis leaves. Frontiers in Plant Science. 2012;**3**:114. DOI: 10.3389/fpls.2012.00114

[79] Bredebach M, Matern U, Martens S. Three 2-oxoglutarate-dependent dioxygenase activities of *Equisetum arvense* L. forming flavone and flavonol from (2S)-naringenin. Phytochemistry. 2011;**72**(7):557-563. DOI: 10.1016/j.phytochem.2011.01.036

[80] Lee YJ, Kim JH, Kim BG, Lim Y, Ahn JH. Characterization of flavone synthase I from rice. BMB Reports. 2008;**41**(1):68-71. DOI: https://www.ncbi.nlm.nih.gov/pubmed/18304453

[81] Anzellotti D, Ibrahim RK. Molecular characterization and functional expression of flavonol 6-hydroxylase. BMC Plant Biology. 2004;**4**:20. DOI: 10.1186/1471-2229-4-20

[82] Barz W, Welle R. Biosynthesis and metabolism of isoflavones and pterocarpan phytoalexins in chickpea, soybean and phytopathogenic fungi. In: Stafford HA, Ibrahim RK, editors. Phenolic Metabolism in Plants. Boston, MA: Springer US; 1992. pp. 139-164

[83] Du YG, Chu H, Chu IK, Lo C. CYP93G2 is a flavanone 2-hydroxylase required for C-glycosylflavone biosynthesis in rice. Plant Physiology. 2010;**154**(1):324-333. DOI: 10.1104/pp.110.161042

[84] Britsch L. Purification and characterization of flavone synthase I, a 2-oxoglutarate-dependent desaturase. Archives of Biochemistry and Biophysics. 1990;**282**(1):152-160. DOI: 10.1016/0003-9861(90)90099-K

[85] Martens S, Forkmann G. Cloning and expression of flavone synthase II from Gerbera hybrids. The Plant Journal. 1999;**20**(5):611-618. DOI: 10.1046/j.1365-313X.1999.00636.x

[86] Anzellotti D, Ibrahim RK. Novel flavonol 2-oxoglutarate dependent dioxygenase: Affinity purification, characterization, and kinetic properties. Archives of Biochemistry and Biophysics. 2000;**382**(2):161-172. DOI: 10.1006/abbi.2000.2002

[87] Latunde-Dada AO, Cabello-Hurtado F, Czittrich N, Didierjean L, Schopfer C, Hertkorn N, et al. Flavonoid 6-hydroxylase from soybean (*Glycine max* L.), a novel plant P-450 monooxygenase. The Journal of Biological Chemistry. 2001;**276**(3):1688-1695. DOI: 10.1074/jbc.M006277200

[88] Artigot MP, Baes M, Dayde J, Berger M. Expression of flavonoid 6-hydroxylase candidate genes in normal and mutant soybean genotypes for glycitein content. Molecular Biology Reports. 2013;**40**(7):4361-4369. DOI: 10.1007/s11033-013-2526-2

[89] Chebil L, Anthoni J, Humeau C, Gerardin C, Engasser JM, Ghoul M. Enzymatic acylation of flavonoids: Effect of the nature of the substrate, origin of lipase, and operating conditions on conversion yield and regioselectivity. Journal of Agricultural and Food Chemistry. 2007;**55**(23):9496-9502. DOI: 10.1021/jf071943j

[90] Hirschmann F, Krause F, Papenbrock J. The multi-protein family of sulfotransferases in plants: Composition, occurrence, substrate

specificity, and functions. Frontiers in Plant Science. 2014;**5**:556. DOI: 10.3389/fpls.2014.00556

[91] Gidda SK, Varin L. Biochemical and molecular characterization of flavonoid 7-sulfotransferase from *Arabidopsis thaliana*. Plant Physiology and Biochemistry. 2006;**44**(11-12):628-636. DOI: 10.1016/j.plaphy.2006.10.004

[92] Sasaki K, Tsurumaru Y, Yamamoto H, Yazaki K. Molecular characterization of a membrane-bound prenyltransferase specific for isoflavone from *Sophora flavescens*. The Journal of Biological Chemistry. 2011;**286**(27):24125-24134. DOI: 10.1074/jbc.M111.244426

[93] Sasaki K, Mito K, Ohara K, Yamamoto H, Yazaki K. Cloning and characterization of naringenin 8-prenyltransferase, a flavonoid-specific prenyltransferase of *Sophora flavescens*. Plant Physiology. 2008;**146**(3): 1075-1084. DOI: 10.1104/pp.107.110544

[94] Liu Y, Gao L, Liu L, Yang Q, Lu Z, Nie Z, et al. Purification and characterization of a novel galloyltransferase involved in catechin galloylation in the tea plant (*Camellia sinensis*). The Journal of Biological Chemistry. 2012;**287**(53):44406-44417. DOI: 10.1074/jbc.M112.403071

[95] Zhang HH, Yu WY, Li L, Wu F, Chen Q, Yang Y, et al. Protective effects of diketopiperazines from Moslae Herba against influenza A virus-induced pulmonary inflammation via inhibition of viral replication and platelets aggregation. Journal of Ethnopharmacology. 2018;**215**:156-166. DOI: 10.1016/j.jep.2018.01.005

[96] Jeong EJ, Yang H, Kim SH, Kang SY, Sung SH, Kim YC. Inhibitory constituents of *Euonymus alatus* leaves and twigs on nitric oxide production in BV2 microglia cells. Food and Chemical Toxicology. 2011;**49**(6):1394-1398. DOI: 10.1016/j.fct.2011.03.028

[97] Chen L, Ding L, Yu A, Yang R, Wang X, Li J, et al. Continuous determination of total flavonoids in *Platycladus orientalis* (L.) Franco by dynamic microwave-assisted extraction coupled with on-line derivatization and ultraviolet-visible detection. Analytica Chimica Acta. 2007;**596**(1):164-170. DOI: 10.1016/j.aca.2007.05.063

[98] Chen S, Xing XH, Huang JJ, Xu MS. Enzyme-assisted extraction of flavonoids from *Ginkgo biloba* leaves: Improvement effect of flavonol transglycosylation catalyzed by *Penicillium decumbens* cellulase. Enzyme and Microbial Technology. 2011;**48**(1):100-105. DOI: 10.1016/j.enzmictec.2010.09.017

[99] Peng S, Peng MJ, Huang ME, Bu XY, Wu G. Extraction and separation of flavonoids from leave of *Eucommia ulmoides* in aqueous two-phase system. Zhong Yao Cai. 2009;**32**(11):1754-1757. DOI: 10.13863/j.issn1001-4454.2009.11.039

[100] Fu XQ, Ma N, Sun WP, Dang YY. Microwave and enzyme co-assisted aqueous two-phase extraction of polyphenol and lutein from marigold (*Tagetes erecta* L.) flower. Industrial Crop Production. 2018;**123**:296-302. DOI: 10.1016/j.indcrop.2018.06.087

[101] Tang LJ, Zhang SF, Yang JZ, Gao WT. New synthetic methods of flavones. Chinese Journal of Organic Chemistry. 2004;**24**(8):882-889

[102] Lu YH, Ji Z, Qi JX, Du CP, Chen RC, Wu SC. Synthesis of luteolin and kaempferol (author's transl). Yao Xue Xue Bao. 1980;**15**(8):477-481. DOI: 10.16438/j.0513-4870.1980.08.005

[103] Zhu S, Wu J, Du G, Zhou J, Chen J. Efficient synthesis of eriodictyol from L-tyrosine in *Escherichia coli*. Applied and Environmental Microbiology. 2014;**80**(10):3072-3080. DOI: 10.1128/AEM.03986-13

[104] Trantas E, Panopoulos N, Ververidis F. Metabolic engineering of the complete pathway leading to heterologous biosynthesis of various flavonoids and stilbenoids in *Saccharomyces cerevisiae*. Metabolic Engineering. 2009;**11**(6):355-366. DOI: 10.1016/j.ymben.2009.07.004

[105] Miyahisa I, Funa N, Ohnishi Y, Martens S, Moriguchi T, Horinouchi S. Combinatorial biosynthesis of flavones and flavonols in *Escherichia coli*. Applied Microbiology and Biotechnology. 2006;**71**(1):53-58. DOI: 10.1007/s00253-005-0116-5

[106] Koopman F, Beekwilder J, Crimi B, van Houwelingen A, Hall RD, Bosch D, et al. De novo production of the flavonoid naringenin in engineered *Saccharomyces cerevisiae*. Microbial Cell Factories. 2012;**11**:155. DOI: 10.1186/1475-2859-11-155

[107] Duan L, Ding W, Liu X, Cheng X, Cai J, Hua E, et al. Biosynthesis and engineering of kaempferol in *Saccharomyces cerevisiae*. Microbial Cell Factories. 2017;**16**(1):165. DOI: 10.1186/s12934-017-0774-x

[108] Xu K, Lv B, Li C. Cell-free synthetic biotechnology—Multi-enzyme catalysis and biosynthesis. Scientia Sinica Chimica. 2015;**45**(5):429-437. DOI: 10.1360/N032014-00263

[109] Nelson DR, Kamataki T, Waxman DJ, Guengerich FP, Estabrook RW, Feyereisen R, et al. The P450 superfamily: Update on new sequences, gene mapping, accession numbers, early trivial names of enzymes, and nomenclature. DNA and Cell Biology. 1993;**12**(1):1-51. DOI: 10.1089/dna.1993.12.1

[110] Leonard E, Yan Y, Koffas MA. Functional expression of a P450 flavonoid hydroxylase for the biosynthesis of plant-specific hydroxylated flavonols in *Escherichia coli*. Metabolic Engineering. 2006;**8**(2):172-181. DOI: 10.1016/j.ymben.2005.11.001